复旦卓越规划·21世纪大学农业英语系列

农业学术英语文献阅读教程

Academic English for Agriculture: A Literature Reading Coursebook

主 编 姜 梅 张吟松

副主编 李青芮 马 娜 吴广平 王晓蕾

编 委（按姓氏拼音排序）

郭小丽 姜 梅 李青芮 罗递敏

马 娜 王晓蕾 吴广平 徐旭艳

杨 皓 余 珊 张莉诗 张吟松

复旦大學出版社

前　言

2022年12月底，习近平同志在中央农村工作会议上强调，要依靠科技和改革双轮驱动加快建设农业强国。要紧盯世界农业科技前沿，大力提升我国农业科技水平，加快实现高水平农业科技自立自强。要着力提升创新体系整体效能，解决好各自为战、低水平重复、转化率不高等突出问题。要打造国家农业科技战略力量，支持农业领域重大创新平台建设。云南农业大学作为一所农业院校，围绕农业，扎根云南，面向世界，服务全国，我们的职能就是在中央大政方针的指引下、通过对学生进行系统的课程教学和实践操作，使其具备掌握现代化农业技术和管理方法的能力，为推动中国农业现代化建设、加快建设农业强国提供人才保障。本书着眼于农业院校学生在农业科学领域的学术英语交流能力的培养，指导学习者研读英文农业学术文献，了解和掌握学术文献的基本构成要素、篇章结构和语言特征，为将来用英语在国际舞台上交流和传播农业研究成果打下基础。

一、教材定位及编写理念

《农业学术英语文献阅读教程》面向农科专业的高年级本科生、硕士研究生、博士研究生，以及其他农科英语爱好者，既突出了农科的专业性，也强调了学术英语的篇章和语言特点。在选材上，我们参考了《新农科人才培养引导性专业指南》，力求让素材充分体现农业学科特色，涵盖农学、食品、烟草、资环、兽医、植保、园林园艺等农科领域。此外，我们参考了由Olwyn Alexander、Sue Argent、Jenifer Spencer所著书籍*EAP Essentials — A Teacher's Guide to Principles and Practice*主张的语类教学法，紧扣学术英语的语类特点，进行教材内容的布局与素材的分析。本教材旨在让学习者通过学习本课程，既能了解本专业的学科知识，又能通过农业学术英语文献阅读和分析等语言输入活动，提高学生的学术英语阅读能力和写作输出能力。

二、教材内容

英文学术文献作为一种特定的语类，其遣词造句、语篇结构、语言风格都有自己的特点，与一般的日常语篇区别很大。教材主要包括学术文献的组成部分以及学术篇章的结构和语言特点两大部分，重点在于帮助学生提升在学术语境下进行深入阅读和学术探索的能力。教材所用素材大都改编自国内外农业方面的权威期刊英语论文，语言符合学术规范，内容聚焦农业热点。通过对这类素材进行深入解读和分析，帮助学生了解农业类学术论文的逻辑结构、句式特点、行文规范以及学术词汇，掌握阅读策略，提高思辨能力和学术英语论文的写作能力。

教材共有七个单元，Unit 1介绍了研究论文的主要结构和特点，Unit 2至Unit 7分别介绍论文的摘要、引言、方法、结果、讨论、结论的语步（moves）或要素（elements）、常用的句型（sentence structures frequently used）、学术文献的篇章结构和语言特点（organization、rhetorical function、register和cohesion）。各个知识点都配有例文及解析，其后都配有大量练习题，便于学生理解、掌握和应用相关知识点。在每个单元的最后还增加了阅读理解题和词汇练习题，旨在让学生在巩固本单元所学知识的同时，掌握一些常用的学术词汇。

Unit 2至Unit 7具体包含以下内容：

1) Subgenre (Abstract, Introduction, Methods, Results, Discussion, Conclusion);

2) Organization (General-specific; Specific-general);

3) Rhetorical Function (Definition; Problem and Solution; Process; Graph Description; Cause and Effect, Comparison and Contrast; Evidence and Conclusion);

4) Register (Noun Phrase; Modification; Tense in Citation; Passive Voice; Long Sentences; Hedging; Nominalization);

5) Cohesion (Conjunctions, Reference, General Nouns, Substitution, Lexical Reiteration, Lexical Chains);

6) Assignments (Reading Comprehension and Vocabulary).

三、学习指导建议

为了让学生充分掌握学习内容，教师可对每一章节的知识点先进行讲解，然后让学生通过练习巩固所学知识点。通过讲练结合，帮助学生掌握农科学术论文写作的基本要素、篇章结构和语言表达。因此，在教学中，我们建议教师在讲解知识点的同时要重视对学生作业的反馈，在讲解教材内容的同时要让学生结合自己的专业领域的学术论文进行分析和讨论，并进行写作实践。

四、编写团队

姜梅负责本书的总体构思、各单元的设计、第二单元的编写、本书的统稿、初审和第三轮审稿；张吟松召集有关编务会议，拟定各板块内容的编辑计划，负责本书的总体构思、初审和终审；李青芮负责本书的总体构思、各单元的设计、本书第一单元的初审以及本书的第二轮审稿；马娜负责第一单元的编写，第二单元和第三单元的初审；吴广平负责第三单元的编写，第四单元和第六单元的初审；王晓蕾负责第六单元的编写，第五单元和第七单元的初审；余珊负责第四单元的编写；罗递敏负责第五单元的编写；徐旭艳负责第七单元的编写；张莉诗协助完成第二单元和第三单元的编写，完成各单元Subgenre和Organization部分的第一轮审稿；郭小丽协助完成第四单元和第五单元的编写，完成各单元Rhetorical Function和Register部分的第一轮审稿；杨皓协助完成第六单元和第七单元的编写，完成各单元Cohesion和Assignments部分的第一轮审稿。

本教材的编写团队成员都是长期从事农科研究生学术英语教学的一线老师，虽然有着丰富的农科学术英语教学和实践经验，但由于农科专业知识相对不足，本教材难免存在不成熟和不完善之处，敬请专家、老师和广大读者批评指正。

本书的编写得到了复旦大学出版社的大力支持和帮助，在此要感谢复旦大学出版社，还要特别感谢复旦大学出版社外语分社社长唐敏对本书的设计和内容提出的宝贵意见，以及编辑方君为本书的出版所做的大量辛勤工作。

CONTENTS

Unit 1 Overview of Academic Papers 1

Unit 2 Abstract 13

- **2A Subgenre:** ***Abstract*** 15
- **2B Organization:** ***General-specific (GS)*** 21
- **2C Rhetorical Function:** ***Definition*** 25
- **2D Register:** ***Noun Phrase; Modification*** 28
- **2E Cohesion (1):** ***Conjunctions*** 36
- **2F Assignments:** ***Reading Comprehension and Vocabulary*** 44

Unit 3 Introduction 49

- **3A Subgenre:** ***Introduction*** 51
- **3B Organization:** ***General-specific (GS)*** 63
- **3C Rhetorical Function:** ***Problem and Solution*** 68
- **3D Register:** ***Tense in Citation*** 75
- **3E Cohesion (2):** ***Reference*** 79
- **3F Assignments:** ***Reading Comprehension and Vocabulary*** 82

Unit 4 Methods 85

- **4A Subgenre:** ***Methods*** 87
- **4B Organization:** ***General-specific*** 95
- **4C Rhetorical Function:** ***Process*** 98
- **4D Register:** ***Passive Voice*** 102
- **4E Cohesion (3):** ***General Nouns*** 105
- **4F Assignments:** ***Reading Comprehension and Vocabulary*** 109

Unit 5 Results 115

5A **Subgenre:** ***Results*** 117

5B **Organization:** ***General-specific (GS)*** 123

5C **Rhetorical Function:** ***Graph Description*** 126

5D **Register:** ***Long Sentences*** 132

5E **Cohesion (4):** ***Substitution*** 137

5F **Assignments:** ***Reading Comprehension and Vocabulary*** 144

Unit 6 Discussion 149

6A **Subgenre:** ***Discussion*** 151

6B **Organization:** ***Specific-general (SG)*** 164

6C **Rhetorical Function:** ***Cause and Effect; Comparison and Contrast*** 170

6D **Register:** ***Hedging*** 177

6E **Cohesion (5):** ***Lexical Reiteration*** 182

6F **Assignments:** ***Reading Comprehension and Vocabulary*** 186

Unit 7 Conclusion 193

7A **Subgenre:** ***Conclusion*** 195

7B **Organization:** ***General-specific (GS)*** 202

7C **Rhetorical functions:** ***Evidence and Conclusions*** 207

7D **Register:** ***Nominalization*** 211

7E **Cohesion (6):** ***Lexical Chains*** 216

7F **Assignments:** ***Reading Comprehension and Vocabulary*** 220

Appendix: Articles Used as Examples 225

References 229

Unit 1 Overview of Academic Papers

1. A Brief Introduction to Academic Papers

Academic papers are professional works usually published in academic journals. They call for wider research through thinking and planning, constructing theme, wording and phrasing, reasoning and proving. They are usually formal, substantial and well-documented, may contain original research results or may be a review of existing research results, and typically fall into a particular category such as research papers or review papers.

A research paper analyzes a perspective or argues a point, and it presents the author's own interpretation, evaluation or argument, backed up by others' ideas and information, more than the sum of sources, or a collection of different pieces of information about a topic. Unlike theses or dissertations, research articles are characterized by novelty or originality.

A review paper analyzes or discusses research previously published by others, rather than reporting new experimental results. Review paper is sometimes called survey overview paper, which mainly concern recent major advances and discoveries, significant gaps in the research, current debates, and possible future research.

2. Key Features of Academic Papers

An academic paper, an academic journal genre like other genres, is conventionally organized in terms of particular textual structures, rhetorical features and lexical/grammatical forms, which also varies according to different purposes and audience (Swales, 1990; Nesi & Gardner, 2012). As the table (Alexander, Argent & Spencer, 2008) on next page shows, there are five distinctive features of academic papers.

Table 1.1 Key Features of Academic Papers

Genre	the types of texts used by groups who share communicative purposes, e.g. case study, research paper
Organization	the development in academic texts from general to specific or from specific to general
Rhetorical Function	a set of rules that guide a writer in creating an effective composition
Register	the style of language used in a particular context, e.g. formal or informal
Cohesion	the ways in which text is tied together: structural, e.g. thematic progression; lexical, e.g. summarizing nouns

2.1 Genre

Academic papers are products of communicative events aimed at specific audiences and intended to achieve particular purposes. In order to achieve their overall purpose, genres proceed through a number of moves and steps, which has a specific rhetorical objective. These moves express the intention of the writer at particular points in the text and do not map precisely on sentences or paragraphs. The structures of academic papers vary from discipline to discipline, journal to journal, and even from research paradigm to research paradigm (Braine, 1995). The structures also vary in terms of the purpose (e.g. testing hypotheses, solving problems, providing theorems) and the mode of inquiry (logical deduction, experiment, case study or interview). The diagrams in the following figure (Cai, 2020) show the conventional structure formats of academic papers.

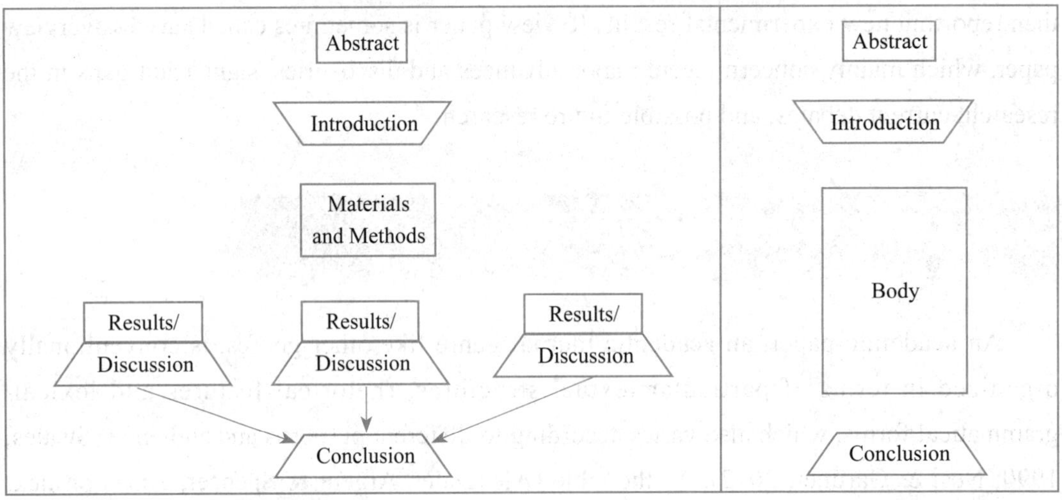

(To be continued)

(Continued)

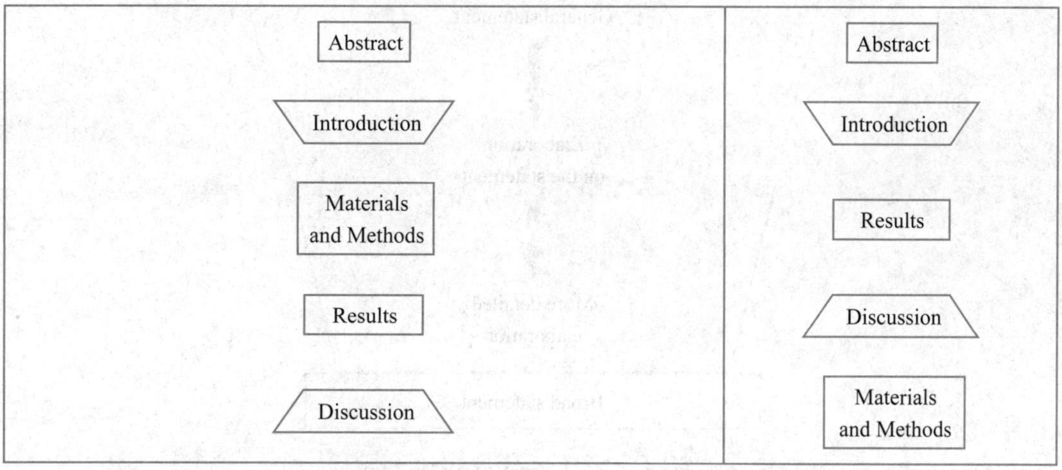

Figure 1.1 Different Structures in Writing an Academic Paper

Each part of a certain academic structure is a sub-genre. An A-IMRD model, a common format in academic writing, includes the sub-genres: Abstract, Introduction, Methods, Results and Discussion. Each has its unique moves, steps and language characteristics, and those will be discussed in detail in the following units.

2.2 Organization

Although variations (e.g. A:RMID) exist across different journals and disciplines, most share the "general-specific pattern", which means that an academic paper usually starts with general information of the research area, and widens in the body part of methods and result, and finally narrows down in the conclusion part.

General-specific (GS) texts are quite common in graduate students' writing, and they are comparatively straightforward. The GS pattern can be used at the paragraph level as well as for larger units of discourse, such as a series of paragraphs in a section or even the text as a whole.

GS texts typically begin with one of the following: a short or extended definition, a generalization or purpose statement, a statement of fact, or some interesting statistics. While the overall movement is from more general to specific, these texts can widen out again in the final sentence. As shown in Figure 1.2 on the next page (Swales & Feak, 2012), the shape is similar to that of a glass or funnel with a base.

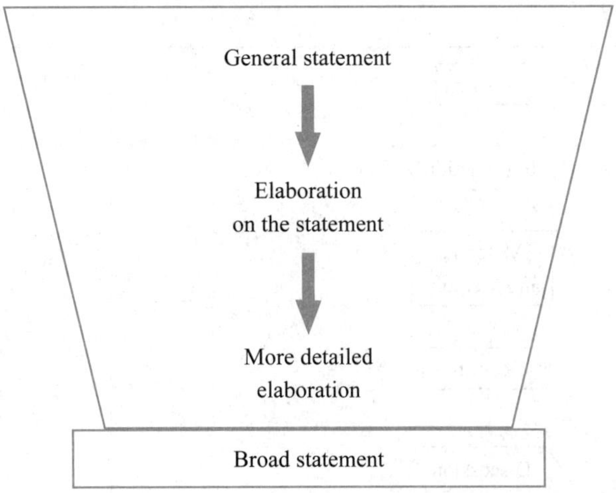

Figure 1.2 Shape of a GS Text

In some cases, the paragraph may be arranged in a specific-general pattern, presenting the information in an inductive way. This pattern gradually generalizes what has been found and involves the readers in a process of discovery, thus arousing their interests and curiosity. While most texts are developed in a simpler GS mode that presents the topic sentence at the very beginning, SG paragraphs serve as a variation that requires the reader to read more actively.

2.3 Rhetorical function

As shown in Figure 1.3 (Alexander, Argent & Spenser, 2008), when writers want to show relationships between ideas, they use rhetorical functions such as comparing, defining, classifying. These functions operate at the paragraph level and influence the choice of vocabulary and sentence structure. They could be thought of as the basic building blocks of texts. Teachers often use rhetorical functions to give students feedback on their writing, such as poorly defined terms, confusion between cause and effect, inexplicit explanation of the problem and weak argumentation.

Rhetorical functions can be classified into three main macro-functions, depending on whether the rhetorical purpose is to describe, to explain or to persuade. Within each of these broad categories, the more specific functions which relate to them are listed. There is a progression from knowledge telling, in which the writer presents information in a straightforward way, to knowledge transforming, in which the writer restructures information in order to explain or persuade a reader.

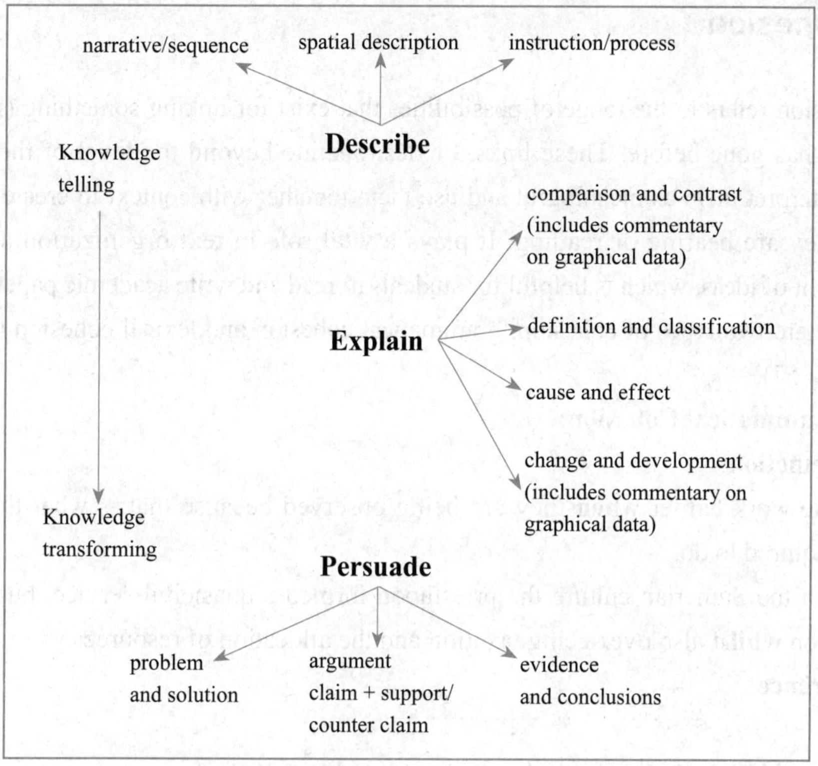

Figure 1.3 Shape of the Rhetorical Functions of Academic Paper

2.4 Register

Register refers to the use of different styles of language in different situations. Language use changes from one context to another depending on the topic under consideration (field), the relationship between the people involved (tenor), and the function of language in the event (mode).

The English language used in an academic paper presents features that are unique enough to establish its own style. The following are some language points to be discussed in this book based on the distinctive linguistic features of academic English:

- noun phrases and modification
- tense in citation
- passive voice
- long sentences
- hedging
- nominalization

2.5 Cohesion

Cohesion refers to the range of possibilities that exist for linking something (in the text) with what has gone before. These links, or ties, operate beyond the level of the sentence. Readers interpret them as meaningful and use them together with context to create coherence in what they are hearing or reading. It plays a vital role in text organization and logical development of ideas, which is helpful for students to read and write academic papers.

There are two types of cohesion: grammatical cohesion and lexical cohesion (Alexander et al., 2008: 57).

★ Grammatical Cohesion:

Conjunction

People work harder **when** they are being observed because that is what they believe they are required to do.

Within the Sumerian culture the priesthood formed a quasicivil service, busying itself with religion **whilst** also overseeing taxation and the allocation of resources.

Reference

Linna told **Hossein** the **secret** but **he** kept **it** to **himself**.

The Internet is useful for finding **information which** is not available in other sources.

Substitution

Dark **objects** absorb more light; lighter **ones** reflect more light.

Information is **linked** into the program automatically if you tell the compiler to **do so**.

Ellipsis

Bacteria of the genus Vibrio are Gram-negative, **[they are]** motile by one or more polar flagella, **[they]** grow on thiosulfate citrate bile salt sucrose agar.

Vibrios are very common in marine and estuarine habitats and **[very common]** on the surface of marine animals.

★ Lexical Cohesion:

Lexical reiteration: repetition, synonym, superordinate

I like drinking **coffee** because **coffee** can keep me awake when I am working.

For better or for worse, **human behavior** has a large influence on global ecology. Many of the environmental challenges facing us today are a direct result of **human actions**, and as such may require behavioral solutions.

Among the **animals**, I like **cats** most.

Lexical chains

Buckle your **seatbelt** for a **ride** on the information **superhighway**.

General nouns

We travelled by train to Edinburgh. The **journey** lasted four and a half hours.

3. Vocabulary in Academic Papers

There are different types of vocabulary used in academic contexts, namely, technical, semi-technical, and academic vocabulary (see Table 1.2 below). In this book, academic vocabulary is mainly discussed in the assignment section of each unit.

Table 1.2 Categories of academic vocabulary

Types of vocabulary	Explanation
Technical	terms specific to a discipline or a particular area of a discipline, which would not usually be understood by a non-specialist outside that discipline
Semi-technical	words or phrases in general use that also have a restricted or special meaning in a particular discipline, or that have different meanings in different disciplines
Academic vocabulary	vocabulary that represents academic activities or academic register, or which signals rhetorical functions or text organization

A Sample Paper

Compatibility of *Amblyseius swirskii* with *Beauveria bassiana:* Two Potentially Complimentary Biocontrol Agents	
Abstract The two biocontrol agents *Amblyseius swirskii* Athias-Henriot (Acari: Phytoseiidae) and *Beauveria bassiana* (Balsamo) Vuillemin (Hypocreales: Cordycipitaceae) have the potential to complement one another as part of an integrated pest management programme. While both can suppress whitefly and thrips infestations in protected crops, *A. swirskii* is primarily used preventatively, whereas *B. bassiana* can be used as a curative treatment at higher pest levels. With this concomitant use in mind, the research presented here aimed to identify potential negative effects of the commercial *B. bassiana* strain GHA on *A. swirskii*	**Subgenre:** Abstract

(To be continued)

(Continued)

Compatibility of *Amblyseius swirskii* with *Beauveria bassiana:* Two Potentially Complimentary Biocontrol Agents	
in a laboratory study. Adult *A. swirskii* were found to be susceptible to *B. bassiana* infection with slight to moderate virulence (20.74 ± 3.89 to 48.33 % ± 3.07) depending on the type of exposure and with a reduction in fecundity over time. There was, however, no negative effect on juvenile survival neither on dry residue nor on the offspring of infected mites. Thus, these two biocontrol agents do have the potential to be complementary. Further trials in the field are still required before a final conclusion can be reached.	
Introduction Augmentative biocontrol in protected crops often involves multiple biological control agents (BCAs) targeting the same or different pest species (Wittman and Leather 1997; Shipp et al. 2003; Calvo et al. 2009; Labbe et al. 2009; Chow et al. 2010; Messelink et al. 2011). These BCAs may also be used in conjunction with biorational chemicals in integrated pest management (IPM) programmes where the focus is to combine crop protection strategies with the overall aim of reducing the use of broad spectrum pesticides and chemical residues on produce (Cock et al. 2010). Combining several BCAs may allow a holistic biological pest management programme, but interactions between different BCAs are complex and one species can potentially disrupt the functions of another. Biocontrol programmes can be compromised due to competition or intraguild predation (e.g. hyperpredation, hyperparasitism and pathogenicity) from other introduced BCAs or naturally occurring species (Momen and Abdel-Khalek 2009; Chow et al. 2010; Buitenhuis et al. 2010; Messelink et al. 2011; Shipp et al. 2012; da Silva et al. 2015). In order to develop efficient and successful pest management programmes, compatibility between the different BCAs must be established. Phytoseiid mites are important components of pest management programmes in protected crops and one of the most widely used species is *Amblyseius* (= *Typhlodromips*) *swirskii* Athias-Henriot (Acari: Phytoseiidae) (van Lenteren 2012). Marketed primarily for whitefly (Hemiptera: Aleyrodidae) and thrips (Thysanoptera: Thripidae) control, both of which are major pests in tomatoes, peppers, cucumbers and ornamentals (Malais and Ravensberg 2003), this polyphagous predator can suppress pest outbreaks on its own (Messelink et al. 2006; Chow et al. 2010; Messelink et al. 2010; Calvo et al. 2011). Combining its use with other BCAs can enhance pest control, particularly when the BCAs do not compete directly but exploit different life stages of the same pest (Calvo et al. 2009; Dogramaci et al. 2011). Conversely, *A. swirskii* can interfere with other BCAs by inter- and intraguild predation (hyperpredation) (Buitenhuis et al. 2010; Messelink et al. 2011). In the field, *A. swirskii* is primarily used preventatively to avert pest establishment and for controlling light whitefly and thrips infestations (Messelink et al. 2010; Calvo et al. 2011). At high pest pressure, curative control by predatory mites is often insufficient and corrective applications with chemical	**Subgenre:** Introduction

(To be continued)

(Continued)

<table>
<tr><th>Compatibility of Amblyseius swirskii with Beauveria bassiana: Two Potentially Complimentary Biocontrol Agents</th><th></th></tr>
<tr><td>pesticides or biopesticides may be required (Calvo et al. 2009; Medd and GreatRex 2014). Ideally, the products used for corrective treatments should have minimal impact on other BCAs present in the crop.

The fungal entomopathogen B. bassiana (Balsamo) Vuillemin (Hypocreales: Cordycipitaceae) exhibits a wide host range and several strains have been developed as commercial mycoinsecticides (de Faria and Wraight 2007; Zimmermann 2007). Commercial formulations of B. bassiana are used curatively to suppress larger outbreaks of arthropod pests (Wraight et al. 2000; Faria and Wraight 2001; Inglis et al. 2001; Jacobson et al. 2001; Shipp et al. 2003). As both A. swirskii and B. bassiana target whitefly and thrips, it is natural to consider them as potentially complementary intraguild BCAs: A. swirskii for preventative control and B. bassiana for curative control.

Due to its wide host range, B. bassiana may have detrimental effects on beneficial arthropods but this will depend on pathogen strain and host species (Seiedy et al. 2015). For instance, B. bassiana has been found to be infectious to hymenopteran parasitoids (Ludwig and Oetting 2001; Shipp et al. 2003) and may potentially disrupt their biocontrol functions. Despite the wide host range of B. bassiana, pathogenicity has only been reported to relatively few species of mites (Chandler et al. 2000).

Interactions between arthropods and entomopathogens are however complex (Agboton et al. 2013; Aqueel and Leather 2013) and the impact of sublethal effects is poorly understood (Pozzebon and Duso 2010; Seiedy et al. 2012; Shipp et al. 2012). The aim of this study was to establish the potential of the commercially available B. bassiana strain GHA to infect and kill A. swirskii under optimal conditions for the pathogen and to investigate the effects on specific population parameters when subjected to topical application of B. bassiana conidial suspensions and/or dry residues on leaf discs.</td><td></td></tr>
<tr><td>Materials and methods
Mite cultures
A. swirskii were acquired from a commercial culture at BCP Certis (Kent, UK) and reared on 50 mm French dwarf bean, Phaseolus vulgaris L. leaf discs at 25 °C, 70% RH and 16:8 L:D. Cattail pollen (Typha sp.) was provided as a food source ad libitum.
Fungal cultures
Laboratory cultures of B. bassiana strain GHA were established from the commercial product BotaniGard® 22 WP (Laverlam International Corporation, USA) by spread-plating 100 μl of a 0.0625% suspension onto Sabouraud Dextrose Agar (SDA) in 90 mm diameter Petri dishes. Cultures were incubated in darkness at 25 °C and 75% RH for 14 days. Sub-cultures were prepared by spread-plating 100 μl conidial suspensions.
Preparation of conidial suspensions
Following 14 days of incubation, the B. bassiana cultures were dried by removing</td><td>Subgenre:
Materials and methods</td></tr>
</table>

(To be continued)

(Continued)

Compatibility of *Amblyseius swirskii* with *Beauveria bassiana:* Two Potentially Complimentary Biocontrol Agents	
the Petri dish lid and placing the cultures in 25 °C and 40% RH for 18 h (Baxter 2008). Conidia were dislodged and collected by tapping the Petri dish containing the culture into a sterile glass funnel (100 mm diameter) leading to a sterile 50 ml glass bottle. The conidia collected from 10 plates were suspended in 5 ml sterile water with 0.02% Tween® 80, and vortex mixed for 2 min (Mascarin et al. 2013). Conidial suspensions were prepared from dried cultures to represent the conidia from the dry formulation of the commercial product. The concentration of conidia was assessed using a Neubauer Haemocytometer 0.1 mm (Hawksley, UK). Conidial suspensions were diluted to 2.5×10^7 and 1×10^8 conidia ml^{-1}, representing the lowest and highest recommended rates, for whitefly and thrips respectively, according to the BotaniGard® 22 WP label. Six replicate conidial suspensions were prepared per treatment for each experiment. To estimate viability of the conidia, 10 μl of the conidial suspension was applied onto a thin layer of SDA on one microscope slide per replicate conidial suspension (n = 6) placing a cover slip on top (Bugeme et al. 2008). The slides were incubated in darkness at 25 °C, 75% RH for 18 h. Germination rate was estimated by assessing 100 conidia in four different fields of view per slide (Liu et al. 2003). Viability was $>$ 95 % in all experiments. Bioassays Female adults of *A. swirskii*, age 2–4 days, were exposed to *B. bassiana* by topical application, dry residue on *P. vulgaris* leaf discs or a combined exposure with topical application followed by dry residue exposure, inspired by the methods of Dennehy et al. (1993) and Pozzebon and Duso (2010) as described below. The trial arenas (leaf discs in 50 mm Petri dishes) were incubated at 25 °C, 90% RH and 16:8 L:D to maximise the likelihood of fungal infection. Topical application of *A. swirskii* females was conducted by immersing 10 mites individually in 1 ml of *B. bassiana* conidial suspensions of 2.5×10^7 or 1×10^8 conidia ml^{-1} in a Petri dish for 30 s. The mites were then placed on filter paper to drain off excess liquid prior to transfer onto leaf discs using a fine brush. This procedure was replicated six times, hence 60 mites were tested per treatment. Control mites were immersed in a 0.02% Tween® 80 solution. Pollen (*Typha* sp.) was provided as a food source in all treatments throughout the experiments. Dry residues of *B. bassiana* were prepared by immersing 50 mm leaf discs in 10 ml conidial suspensions of 2.5×10^7 or 1×10^8 conidia ml^{-1}, and 0.02% Tween® 80 for the control, for 30 s and air drying them for 1 h. The mites in the dry residue and combined exposure trials were allowed to forage on *B. bassiana* residues for four days (Bugeme et al. 2008), representing the persistence of conidia on the top leaves of a plant in the field (Inglis et al. 1993). Survivorship and fecundity were monitored once per day for 10 days, removing the eggs from the leaf discs each day. On day 1, 4, 7 and 10 post-treatment, eggs were	

(To be continued)

(Continued)

<table>
<tr><th>Compatibility of Amblyseius swirskii with Beauveria bassiana: Two Potentially Complimentary Biocontrol Agents</th><th></th></tr>
<tr><td>collected from each leaf disc using a fine brush and placed onto untreated leaf discs. The offspring were monitored for six days recording egg viability, juvenile survival and sex ratio.

Dead mites were transferred to Petri dishes with moistened filter paper to encourage sporulation. Death by mycosis was confirmed by sporulation or red colouration of cadavers. The identity of the fungal growth on cadavers was confirmed by microscopic investigation of lactophenol blue stained conidiophores and conidia at magnification ×400, according to Humber (1997).

The effect of B. bassiana residues on juvenile A. swirskii was studied to simulate the impact on new founding populations of mites dispersing into a newly treated crop. Ten gravid A. swirskii were allowed to oviposit on dry residue of the conidial suspensions, or 0.02% Tween® 80 as a control, for 16 h. The eggs deposited were corrected to 15 per replicate leaf disc (n = 6) and reared to adulthood for six days on the dry residue, recording egg viability and juvenile survival. Six days was sufficient for all mites to have reached maturity and to have mated. The surviving adult females were transferred to clean leaf discs and monitored for a further 10 days for survivorship and fecundity.

Data analysis

Survival rates were estimated by parametric survival analysis with the Weibull distribution, due to non-constant hazards, in the R 3.1.2 statistical package (R Core Team 2012), which generates z values as a result of Wald tests. Multiple comparisons were conducted using Tukey's post-hoc in the multcomp package in R (Hothorn et al. 2008).

...</td><td></td></tr>
<tr><td>Results
Survivorship

The B. bassiana treatments had a significant effect on the survival rates of adult A. swirskii compared with the control treatment. On dry residue the suspension with 1×10^8 conidia ml^{-1} significantly reduced survival compared with the control treatment ($z = 2.97$, $p < 0.01$). There was no significant difference between this treatment and the 2.5×10^7 conidia ml^{-1} treatment ($z = 1.77$, $p = 0.18$), nor between the latter and the control ($z = 1.76$, $p = 0.18$) (Fig. 1a).

Topical application of B. bassiana significantly reduced the survival rates of adult A. swirskii for both 2.5×10^7 ($z = 2.92$, $p < 0.01$) and 1×10^8 conidia ml^{-1} ($z = 3.70$, $p < 0.001$) compared with the control treatment. There was no significant difference ($z = 1.53$, $p = 0.27$) between the two B. bassiana treatments (Fig. 1b).

Combined exposure to topical application and dry residue of B. bassiana significantly reduced the survival rates of adult A. swirskii for both 2.5×10^7 ($z = 3.39$, $p < 0.01$) and 1×10^8 conidia ml^{-1} ($z = 3.68$, $p < 0.001$) compared with the control treatment. There was no significant difference between the two B. bassiana treatments ($z = 0.92$, $p = 0.61$) (Fig. 1c).</td><td>Subgenre:
Results</td></tr>
</table>

(To be continued)

(Continued)

Compatibility of *Amblyseius swirskii* with *Beauveria bassiana:* Two Potentially Complimentary Biocontrol Agents	
B. bassiana did not have a significant effect on survival rates of juvenile *A. swirskii* reared from egg to adult on dry residue and monitored for a further 10 days after final ecdysis (Fig. 1d). There was no treatment effect on egg viability and juvenile survival was similar in all treatments. The scale parameter was <1 in all experiments, indicating that the risk of dying decreases with time (Crawley 2007). This suggests the time frame of observation (10 days) was sufficient to capture the majority of the mites dying as a result of the treatment. Mites with confirmed mycosis were observed in all B. bassiana treatments except with juveniles reared on the dry residue of the conidial suspension with 2.5×10^7 conidia ml^{-1}. Dead mites exhibited pink to dark red colouration after death followed by sporulation. No mycosis was observed in the control treatments. Fecundity The oviposition rate decreased with time in all treatments and the number of eggs deposited per surviving female per day decreased at a higher rate for *A. swirskii* treated with *B. bassiana* than control mites (Fig. 2a–d). Exposing adult *A. swirskii* to dry residue of *B. bassiana* did not have a significant effect on fecundity (Fig. 2a). Juvenile *A. swirskii* exposed to dry residues however exhibited a significant decrease in fecundity in the subsequent mature life stage for 1.0×10^8 ($z = 3.90$, $p < 0.001$) and 2.5×10^7 conidia ml^{-1} ($z = 2.62$, $p < 0.05$) (Fig. 2d). For adult mites exposed to topical and combined treatments there was a significant effect on fecundity at 2.5×10^7 ($z = 2.38$, $p < 0.05$) and 1.0×10^8 conidia ml^{-1} ($z = 4.49$, $p < 0.001$), respectively. Treatments with 1.0×10^8 and 2.5×10^7 conidia ml^{-1} did not exhibit a significant effect on fecundity for topical ($z = 2.27$, $p = 0.06$) (Fig. 2b) and combined ($z = 1.76$, $p = 0.18$) exposure (Fig. 2c) respectively, due to within-group variation in the data. Offspring Egg viability, juvenile survival and sex ratio of the offspring from *B. bassiana*-treated adult *A. swirskii* are summarised in Table 1. As there were no differences in any of the measured parameters between sampling ...	
Discussion *B. bassiana* strain GHA was found to be infectious to *A. swirskii* when exposed to topical application and/or dry residues under laboratory conditions with slight to moderate virulence (according to the IOBC toxicity classes). The topical treatment was intended to simulate the effect on a mite population in the crop directly exposed to a *B. bassiana* application followed by dispersal to untreated leaves or new growth. The dry residue exposure simulated mites dispersing into a recently treated crop whereas the combined exposure simulated the effect on mites exposed directly to the spray application and remaining on the treated leaves. ...	**Subgenre:** Discussion

Unit 2 Abstract

Main Contents	Learning Objectives
Subgenre: *Abstract*	★ Understanding the function of an abstract. ★ Identifying the structure of an abstract.
Organization: *General-specific*	★ Learning about GS pattern abstract. ★ Understanding the abstract better by analyzing its organization.
Rhetorical Function: *Definition*	★ Understanding the structure of a formal sentence definition. ★ Learning to write definitions with propositional phrases, participle phrases and relative clauses.
Register: *Noun Phrase; Modification*	★ Recognizing the structure of noun phrases. ★ Learning to write sentences with noun phrases. ★ Learning to use modification in a sentence.
Cohesion (1): *Conjunctions*	★ Learning about the categories of conjunctions used in paragraphs and texts. ★ Learning to use proper conjunctions in paragraphs and texts.
Assignments: *Reading Comprehension and Vocabulary*	★ Reading: Understanding the abstract better through in-depth reading. ★ Vocabulary: Learning general academic words.

2A Subgenre: *Abstract*

What is an abstract? An abstract is a brief summary of the most important points of a scientific research paper. To be specific, an abstract is a stand-alone statement that briefly conveys the essential information of a paper, article, document or book; presents the objective, methods, results and conclusions of a research project; and has a brief, non-repetitive style.

The abstract of a paper gives a first impression of the paper that follows, allowing the reader to decide whether to continue reading and what to look for if they do. Also, the abstract can help readers quickly understand the main content, especially the objective and findings of the study. The abstract is therefore an extremely important part of an academic paper.

1. The structure of an abstract

Although the requirements for the length and format of a paper vary from discipline to discipline, a typical abstract of an agricultural paper should provide: 1) the background to the study, including the study's purpose; 2) methodology; 3) main findings, including principal conclusions. Therefore, the structure of an abstract is always more or less the same. Typically, it is a single paragraph, containing five moves: Background, Purpose, Methods, Results, Conclusions (abbreviated as "BPMRC"). You may check Table 2.1 (Swales & Feak, 2012) for a more detailed introduction.

Table 2.1 The Structure of an Abstract

Move 1 (Background)	Situating the research (optional)	a) Setting the background knowledge (optional) b) Indicating the problem to be addressed (optional)
Move 2 (Purpose)	Announcing the purpose of the research or the principal activity and its scope	

(To be continued)

(Continued)

Move 3 (Methods)	Describing the methods used	a) Stating the material/subjects b) Describing the procedure (optional) c) Justifying the method (optional)
Move 4 (Results)	Reporting the main findings	
Move 5 (Conclusions)	Providing the conclusions	a) Indicating the improvement and the significance or implications (optional) b) Pointing out the application or further research (optional)

Example 1

Title of the paper: *Food Insecurity and Risk of Poor Health Among US-Born Children of Immigrants*

[Objectives] We investigated the risk of household food insecurity and reported fair or poor health among very young children who were US citizens and whose mothers were immigrants compared with those whose mothers had been born in the United States.

[Methods] Data were obtained from 19,275 mothers (7,216 of whom were immigrants) who were interviewed in hospital-based settings between 1998 and 2005 as part of the Children's Sentinel Nutrition Assessment Program. We examined whether food insecurity mediated the association between immigrant status and child health in relation to length of stay in the United States.

[Results] The risk of fair or poor health was higher among children of recent immigrants than among children of US-born mothers (odds ratio [OR]=1.26; 95% confidence interval [CI]=1.02, 1.55; $P<.03$). Immigrant households were at higher risk of food insecurity than were households with US-born mothers. Newly arrived immigrants were at the highest risk of food insecurity (OR=2.45; 95% CI=2.16, 2.77; $P<.001$). Overall, household food insecurity increased the risk of fair or poor child health (OR=1.74; 95% CI=1.57, 1.93; $P<.001$) and mediated the association between immigrant status and poor child health.

[Conclusions] Children of immigrant mothers are at increased risk of fair or poor health and household food insecurity. Policy interventions addressing food insecurity in immigrant households may promote child health.

Example 2

Title of the paper: *Effects of Dietary Energy on Growth Performance, Rumen Fermentation and Bacterial Community, and Meat Quality of Holstein-Friesians Bulls Slaughtered at Different Ages*

[Objectives] The objective of this study was to evaluate the effects of dietary energy levels on growth performance, rumen fermentation and bacterial community, and meat quality of Holstein-Friesians bulls slaughtered at different ages. **[Methods]** Thirty-six Holstein-Friesians bulls (17 months of age) were divided into a 3 × 3 factorial experiment with three energy levels (LE, ME and HE; metabolizable energy is 10.12, 10.90 and 11.68 MJ/kg, respectively) of diets, and three slaughter ages (20, 23 and 26 months). **[Results]** Results indicated that bulls fed with ME and HE diets had higher dry matter intake, average daily gain, and dressing percentage at 23 or 26 months of age. The ME and HE diets also reduced bacterial diversity, altered relative abundances of bacteria and produced lower concentrations of acetate, but higher butyrate and valerate concentrations in rumen fluid. Increase in dietary energy and slaughter age increased the intramuscular fat (IMF) and water holding capacity. **[Conclusions]** In summary, Holstein-Friesians bulls fed with ME and HE diets, slaughtered at 23 and 26 months of age could be a good choice to produce beef with high IMF. Slaughter age may have less influence than dietary energy in altering fermentation by increasing amylolytic bacteria and decreasing cellulolytic bacteria, and thus, further affecting meat quality.

Task 1 Read the following abstracts and match features with the sentences.

Abstract 1

① Virus vectors carrying host-derived sequence inserts induce silencing of the corresponding genes in infected plants. ② This virus-induced gene silencing (VIGS) is a manifestation of an RNA-mediated defense mechanism that is related to post-transcriptional gene silencing (PTGS) in transgenic plants. ③ Here we describe an infectious cDNA clone of tobacco rattle virus (TRV) that has been modified to facilitate insertion of non-viral sequence and subsequent infection to plants. ④ We show that this vector mediates VIGS of endogenous genes in the absence of virus-induced symptoms. ⑤ Unlike other RNA virus vectors that have been used previously for VIGS, the TRV construct is able to target host RNAs in the growing points of plants. ⑥ These features indicate that the TRV vector will have wide application for gene discovery in plants.

a. conclusion ________ b. background ________

c. results ________ d. research scope ________

Abstract 2

① Rose cultivars with blue flower color are among the most attractive breeding targets in floriculture. ② However, they are difficult to produce due to the low efficiency of transformation systems, interactive effects of hosts and vectors, and lengthy processes. ③ In this study, agroinfiltration-mediated transient expression was investigated as a tool to assess the function of flower color genes and to determine appropriate host cultivars for stable transformation in Rosa hybrida. ④ To induce delphinidin accumulation and consequently to produce blue hue, the petals of 30 rose cultivars were infiltrated with three different expression vectors namely pBIH-35S-CcF3′5′H, pBIH-35S-Del2 and pBIH-35S-Del8, harbouring different sets of flower color genes. ⑤ The results obtained showed that the ectopic expression of the genes was only detected in three cultivars with dark pink petals (i.e. "Purple power", "High & Mora" and "Marina") after 6–8 days. ⑥ The high-performance liquid chromatography analyses confirmed delphinidin accumulation in the infiltrated petals caused by transient expression of CcF3′5′H gene. ⑦ Moreover, there were significant differences in the amounts of delphinidin among the three cultivars infiltrated with the three different expression vectors. ⑧ More specifically, the highest delphinidin content was detected in the cultivar "Purple power" (4.67 μg/g FW), infiltrated with the pBIH-35S-Del2 vector. ⑨ The expression of CcF3′5′H gene in the infiltrated petals was also confirmed by real time PCR. ⑩ In conclusion and based on the findings of present study, the agroinfiltration could be regarded as a reliable method to identify suitable rose cultivars in blue rose flower production programs.

a. conclusion ________ b. background ________

c. results ________ d. purpose ________

e. methods ________

Task 2 Read the following abstract and complete the table.

① Human milk (HM) is the primary source of nutrients and bioactive components that supports the growth and development of infants. ② However, the proteins present in human milk may change depending on the period of lactation. ③ In this light, the objective of the present study was to evaluate the effect of lactation period on HM utilizing a data-independent acquisition (DIA) approach to identify the differences in HM whey protein proteomes. ④ As part of the study, whey proteins of January, February, and June in HM were studied. ⑤ The results identified a total of 1,536 proteins in HM whey proteins of which 114 groups were subunits of differentially expressed proteins as revealed by cluster analysis. ⑥ Protein expression was observed to be affected by the period of lactation with expression levels of plasminogen, thrombospondin-1, and tenascin higher during January, keratin, type I cytoskeletal 9 highest in February, and transcobalamin-1 highest in June. ⑦ The results of this study not only contribute to expanding our understanding of human whey proteome but also provide strong evidence for the nutritional difference of HM during different lactation periods.

Moves	Implied questions	The sentences corresponding to the BPMRC elements or the implied questions
Move 1 (Background)	What do we know about the topic?	
Move 2 (Purpose)	What is the study about?	
Move 3 (Methods)	How was it done?	
Move 4 (Results)	What was discovered?	
Move 5 (Conclusions)	What's the significance of the findings?	

2. Sentence structures frequently used in an abstract

In each move of an abstract, there are some commonly used sentence structures that are helpful in writing an abstract or identifying the elements of an abstract.

Table 2.2 Sentence Structures Frequently Used in an Abstract

Moves	Sentence structures
Move 1 (Background)	... is a problem encountered in ... ... is still an open problem ... ... has raised the question of ... ... has been a popular topic for journal articles ... ... However, ...
Move 2 (Purpose)	This paper studies/presents/explores ... This article evaluates ... The aims of this study were twofold. First, ... In this paper, it will be argued that ... This research attempts to point out some main problems of ... The aim/objective/purpose of this paper/investigation is to ...
Move 3 (Methods)	A ... strategy was chosen in order to evaluate ... Experiments were carried out to explore ... We use a ... model, which ..., to examine ... Using a ... survey, we reexamine ... A ... approach is applied in order to ... In this study, we used ... to determine ...
Move 4 (Results)	Results include that ... The result/analysis indicates/shows/suggests that ... The experiments/examples suggest/support/demonstrate ... It is shown experimentally that ... There was no significant difference in ...
Move 5 (Conclusions)	We conclude/suggest that ... This paper ends with the suggestion that ... It is recommended that ... The results can explain ... and provide a tool for a quantitative analysis of ... These results are significant for ... Further improvements are possible by ...

Task 3 Read the above two abstracts in Task 1 again and find out the language used to express moves which can help you to identify the elements (BPMRC).

Abstracts	Background	Purpose	Methods	Results	Conclusions
Abstract 1					
Abstract 2					

2B Organization: *General-specific (GS)*

An abstract usually follows the GS pattern. As discussed in Unit 1, GS texts typically begin with one of the following: a short or extended definition, a generalization or purpose statement, a statement of fact, or some interesting statistics. An abstract, which usually begins with an introduction to the background, the purpose of the research, or a definition of a key concept in the study, often states a fact, a purpose or a definition before the detailed elaboration. So, in terms of organization, it features the GS pattern.

Example 1

① The main purpose of the present study was to explore the optimum timing for mid-stage nitrogen (N) application to improve fragrant rice performance. ② The field experiment was conducted in 2019 and 2020 with two fragrant rice cultivars, i.e., YNX28 and NG9108, and five timings of mid-stage nitrogen application: at the emergence of the top-fifth-leaf (TL5), the top-fourth-leaf (TL4), the top-third-leaf (TL3), the top-second-leaf (TL2), and the top-first-leaf (TL1) of the main stem. ③ The control condition received no mid-stage nitrogen application (CK). ④ The results revealed that the TL5, TL4, TL3 and TL2 treatments significantly increased grain yield of both varieties in 2019 and 2020, compared with CK. ⑤ Grain yield reached a maximum value under the TL3 treatment of YNX28 and the TL4 treatment of NG9108. ⑥ Moreover, the TL4, TL3 and TL2 treatments significantly increased the brown rice rate, milled rice rate, head rice rate and protein content of both varieties in 2019 and 2020, compared with CK, whereas the TL2 and TL1 treatments of YNX28 and the TL3, TL2 and TL1 treatments of NG9108 significantly increased the chalky rice rate and chalkiness. ⑦ Further, appropriate mid-stage nitrogen application timing also improved the grain 2-AP content due to regulation of the contents of synthesis-related precursors (proline, GABA) and enzymatic activities (ProDH, OAT, P5CS, BADH2). ⑧ Compared with CK, the TL4 and TL3 treatments of YNX28 significantly increased the 2-AP content in the rice grains by 21.8%–27.4%, while the TL5 and TL4 treatments of NG9108 significantly increased the 2-AP content in the rice grains by 15.6%–31.6%. ⑨ These results suggest that the TL3 treatment of YNX28 and the TL4 treatment of NG9108 are the optimum timings of mid-stage nitrogen application for fragrant rice production. ⑩ The optimal mid-stage nitrogen application timing of NG9108 is earlier than that of YNX28.

In the abstract, we can see the GS pattern:

Sentence ① states the purpose.

Sentences ② & ③ present the methods to accomplish the purpose.

Sentences ④ & ⑧ describe the results after using the methods to achieve the purpose.

Sentences ⑨ & ⑩ respond to the purpose.

This is a GS-pattern abstract, which is written with a purpose put at the beginning of the paragraph, and the parts following the purpose are some descriptions regarding the purpose.

Example 2

① This virus-induced gene silencing (VIGS) is a manifestation of an RNA-mediated defense mechanism that is related to post-transcriptional gene silencing (PTGS) in transgenic plants. ② Here we describe an infectious cDNA clone of tobacco rattle virus (TRV) that has been modified to facilitate insertion of non-viral sequence and subsequent infection to plants. ③ We show that this vector mediates VIGS of endogenous genes in the absence of virus-induced symptoms. ④ Unlike other RNA virus vectors that have been used previously for VIGS, the TRV construct is able to target host RNAs in the growing points of plants. ⑤ These features indicate that the TRV vector will have wide application for gene discovery in plants.

This abstract follows the GS pattern:

Sentence ① situates the research.

Sentence ② focuses on the research topic.

Sentences ③ & ④ describe the findings of the research and make a comparison.

Sentence ⑤ indicates the significance of the research.

This GS-pattern abstract starts with a statement of fact, which situates the research. Then in the following parts, the topic, the findings and the significance of this research are described.

Task 1 Read the following abstract and finish the exercises.

① Hypsizygus ulmarius polysaccharide (HUP) is a water-soluble polysaccharide obtained by hot water extraction, followed by precipitation and deproteinization. ② The characteristics of HUP, antioxidant activity and liver protection against alcohol-induced liver damage were studied. ③ Structural characteristics indicate that the HUP is a pyran-type polysaccharide with a molecular weight of 2076 Da. ④ In antioxidant scavenging assay, HUP showed moderate DMPD radical scavenging activity, cupric ion reducing antioxidant capacity and inhibitory effect against lipid peroxidation in a dose-dependent manner. ⑤ Regarding *in vivo* hepatoprotective activity, compared with the ethanol induction group, pre-treatment of low and high doses of HUP significantly reduced the behaviours of serum enzymes, lowered the levels of hepatic oxidative stress markers, restored the levels of biochemical constituents, enhanced the levels of liver and serum enzymatic antioxidants and non-enzymatic antioxidants, and improved the serum lipid levels of alcohol-intoxicated rats. ⑥ The hepatoprotective effect of HUP was comparable to positive control silymarin. ⑦ Besides, HUP pre-treatment significantly normalized the histopathological changes induced by ethanol. ⑧ The results indicate that HUP could be used as a functional food and may protect the biological system from oxidative stress through its antioxidant activity, thus having a significant protective effect on acute alcoholic liver injury.

1) This abstract opens with:

A definition ☐

A generalization or purpose statement ☐

A statement of fact ☐

2) This abstract is organized in the pattern of:

General-specific ☐

Specific-general □

3) The following steps describe the flow of information. Read the steps and, referring to Step 1: Write down the corresponding sentence number in Steps 2–4.

Step 1: Define the key word in the research: Sentence ___①___.

Step 2: Describe the purpose related to the key word defined in Sentence ①: Sentence ________.

Step 3: Depict the results responding to the purpose: Sentence ________.

Step 4: Deduce the conclusion from the results: Sentence ________.

Task 2 The following are sentences from a GS abstract. Restructure it by writing the numbers in the correct order.

① However, the level of WMV did not show any significant increase in doubly- versus singly-infected plants. Single infections of WMV or CGMMV on the same hosts produced only vein clearing, blistering, systemic mosaic or mottling on the upper leaves and similar symptoms developed after double infection.

② Mixed infections of cucurbits by Cucumber green mottle mosaic virus (CGMMV) and Watermelon mosaic virus (WMV) exhibit a synergistic interaction.

③ Moreover, the level of capsid protein from CGMMV increased in mixed infection.

④ It is concluded that co-infections with WMV and CGMMV displayed synergistic interaction which could have epidemiological consequences.

⑤ Watermelon, cucumber and cantaloupe co-infected by the WMV and CGMMV displayed synergistic pathological responses, finally in some cases, progressing to vascular wilt and plant death. Accumulation of CGMMV RNAs in a mixed infection with WMV in some cucurbits was higher than infection with CGMMV alone.

Correct order: ________________

2C Rhetorical Function: *Definition*

Definitions are common in academic papers because writers often use definitions to explain concepts. This unit introduces the structure of a formal sentence definition.

The term *definition* comes from the Latin word *definio*, which means "to limit or bound; to interpret ideas or words in terms of each other; to understand one thing by another". Usually, *definition* refers to an explanation of the meaning of a word or phrase, especially in a dictionary, or the act of stating the meanings of words.

Definitions may simply be short, parenthetical additions to a sentence or perhaps a larger part of a paper. The extent of the definition depends on the purpose of the paper, the level of familiarity your audience has with the subject, and the extent to which there is an agreed-upon definition of the concept.

1. Structure of a sentence definition

The following is a common structure of a formal sentence definition.

term	*is/are*	*a/an class*	*that / (on/in/through/ ...) which*	+	distinguishing detail
A solar cell	is	a device	that/which		converts the energy of sunlight into electric energy.

From the structure, it can be seen the term being defined is first assigned to a class to which it belongs and then distinguished from other terms in the class. The class word is a superordinate — a category word, one level of generality above the term. Some common subordinates, or class words, are *technique*, *method*, *process*, *device*, *system*, and so on.

However, a definition of a term or concept is not always fixed. Some varied structures of definitional sentences are displayed in the table on the next page.

Table 2.3 The Structures of Frequently Used Definitional Sentences

key term *+ notional verb + prepositional phrase*	Agronomy is defined as the scientific study of soil. Agronomy refers to the scientific study of soil.
modifier + noun + relative clause + be + ***key term***	A similar, usually commercial, site where a product is manufactured or cultivated is a farm.
key term *+ be + a class/group + (in, through, by, on ...) which, that, where, when + distinguishing detail*	A food chain is a series of living organisms in which each organism eats the one below. Regenerative agriculture is a system of farming techniques and principles that focuses on the health and regeneration of topsoil.
key term *+ be + noun + modifier (prepositional phrase, participle phrase, adjective phrase)*	A wrench is a metal tool for holding and turning objects. A wrench is a metal tool used for holding and turning objects. A wrench is a metal tool useful for holding and turning objects.

Task 1 Recognize the structure of the following definitional sentences.

a. *key term + notional verb + prepositional phrase*
b. *modifier + noun + relative clause+ be + key term*
c. *key term + be + a class or group + (in, by ...) which, that, where, when + distinguishing detail*
d. *key term + be + noun + modifier (prepositional phrase, participial phrase, adjective phrase)*

______ 1) Organic farming is a type of agriculture that benefits from the recycling and use of natural products.

______ 2) Bioactive peptides have been defined as short protein fragments that may have a positive impact on human health.

______ 3) ACE is an enzyme belonging to the dipeptidyl carboxypeptidase family that catalyses the hydrolysis *in vivo* of the plasmatic peptide angiotensin I in the potent vasoconstrictor angiotensin II.

______ 4) Biscuits are flour confections produced from dough and baked to a very low moisture content within a short period of time to make them flaky and crispy.

______ 5) A high resolution, real-time measurement tool that monitors soil moisture at multiple depths (up to one meter) using a combination of AI algorithms and physical models along with several satellite and weather model data sets is High-Definition Soil Moisture (HD-SM).

Task 2 Read the example and write similar definitions for terms 1)–5).

Model: goal/an objective you set for yourself

A goal can be defined as an objective you set for yourself.

1) organism/an animal or plant that is so small that you cannot see it without using a microscope
2) bacterium/a member of large group of unicellular micro-organisms which have cell walls but lack organelles and an organized nucleus, including some which can cause disease
3) zoologist/a scientist who studies zoology
4) pollution/a form of environmental contamination resulting from human activity
5) tobacco/the dried leaves of the tobacco plant that are used for making cigarettes, smoking in a pipe or chewing

Task 3 Write definitions of the following terms in the list using a prepositional phrase, a participle phrase or a relative clause.

crop	agriculture	soil	plant	rainfall

Task 4 Complete the definitions by inserting an appropriate preposition given in the list.

around	from	in	by

1) Seed is the small hard part produced by a plant, ________ which a new plant can grow.
2) Low levels of straying are crucial, since the process provides a source of novel genes and a mechanism ________ which a location can be repopulated should the fish there disappear.
3) Energy balance is a state ________ which the number of calories eaten equals the number of calories used.
4) An axis is an imaginary line ________ which a body is said to rotate.
5) Photosynthesis is a process ________ which sunlight is used to manufacture carbohydrates from water and carbon dioxide.

2D Register: *Noun Phrase; Modification*

1. Noun phrase

Academic papers need to be concise and information-rich, so noun phrases, which can express complex ideas in a few words, are widely used in academic papers. Noun phrases consist of a head noun, which must always be present, and a number of other elements, all of which are optional. Noun phrases can therefore consist of one word or very many words. For example, the sentence, as shown in the following box, with the underlined word(s) added to the phrase, becomes longer and more informative.

<table>
<tr><td align="right">Preparation (head noun)
The preparation
The preparation of fruits and vegetables
The preparation of processed fruits and vegetables
The preparation of minimally processed fruits and vegetables</td><td>entails physical wounding of the tissue.</td></tr>
</table>

Noun phrases can take many forms. The following are the main structures of noun phrases.

Table 2.4 The Structure of Noun Phrases

determiner and/or enumerator + head noun	an **argument**
determiner and/or enumerator + adjective + head noun	a popular **argument**
determiner and/or enumerator + noun + head noun	the plant **remains** / the countryside **consumption**

(To be continued)

(Continued)

determiner and/or enumerator + adjective + noun + head noun	a rich study **area**
determiner and/or enumerator + head noun + prepositional phrase	the **demands** on the soil
determiner and/or enumerator + head noun + relative clause	the **wheat** that is used to make food
determiner and/or enumerator + head noun + participle phrase	the **research** carried out by the university
determiner and/or enumerator + head noun + prepositional phrase + relative clause	the **writing** about agriculture that was already well-known
determiner and/or enumerator + head noun + adjective phrase	the **food** rich in nutrients

Noun phrases can be structured as the following:

Noun Phrase (NP) = (____ ____ ____ ____)

a b c d

a = determiner and/or enumerator

b = pre-head modification (adjective/noun modifying the head noun)

c = head noun

d = post-head modification (extra information following the head noun, including prepositional phrase, participle phrase, infinitive phrase, adjective phrase, attributive clause, appositive clause)

Examples

1. The protein-protein interaction network was analyzed using STRING.
   a b b c
2. The cut surface of some fruits and vegetables may brown rapidly.
   a b c d (prepositional phrase)
3. One of the latest developments suitable for use by the online marketers.
   a b c d (adjective phrase)
4. This rule is supported by the fact that intronless genes are immune to NMD in mammals.
   a c d (appositive clause)

Task 1 **Read the sentences and identify the parts underlined by writing down the corresponding letter.**

a = determiner and/or enumerator

b = pre-head modification (adjective/noun modifying the head noun)

c = head noun

d = post-head modification (extra information following the head noun, including prepositional phrase, participle phrase, infinitive phrase, adjective phrase, attributive clause, appositive clause)

1) Eight storage trials were conducted.

( )( )( )

2) The transpiration rate through the wounded surface of sweet potato roots was determined

( )( ) ( ) ( )

using a pyrometer.

3) The relationship between L.S. and humidity is examined in more detail in Fig. 2.

( )( ) ( )

4) They reported a thick desiccated layer (17 cell layers) where the roots were kept at 21 °C and

( )( )( ) ( ) ( )

60% RH.

5) The views expressed are not necessarily those of DFID [R6507: Crop Post-Harvest

( )( ) ( )

Research Programme].

2. Modification

Modification (putting modifiers before or after nouns), which can make the articles more concise and succinct, is used widely in academic papers. As we all know like adjectives, participles (*v.*-ing/*v.*-ed) are often used as modifiers, such as "*unhealthy eating habits*", "*the moderating factors that influence HCP weight loss advice*" and "*the observed decrease in soil respiration rates*". Sometimes the modifiers are more complicated, and contain more information, then longer attributives such as subordinate clauses, are required.

Academic writing tends to be organized in such a way that it may pack more information into a fewer words (Biber & Gray, 2016). To make sentences concise and compact, authors tend to use adjective phrases or *v.*-ing/*v.*-ed phrases as modifiers instead of subordinate clauses, especially relative clauses. The following example is the title of an academic paper. The use of two participle phrases helps to keep the title clear, concise and informative as well.

Dose-related ***changes*** *in respiration and enzymatic activities in* ***soils*** *amended with mobile platinum and gold*

The two past participle phrases appear in different places in the title, one before the modified noun and the other after the noun. Generally, shorter phrases can be put before nouns as pre-modifiers (e.g. *a widely accepted concept*). However, if more words are involved (e.g. *a concept that is mostly accepted in Asian countries*), it would be improper to convert the relative clause into a pre-modifier. In this case, the modifying phrases usually follow the noun as a post-modifier (e.g. *a concept mostly accepted in Asian countries*).

Pre-modifier

Modifier phases before the modified nouns usually consist of two or three words, arranged in relatively fixed patterns. Compare the two versions of the following sentences:

1a. Low-income and ethnic minority groups have a greater risk of obesity and illnesses that are related to obesity.

1b. Low-income and ethnic minority groups have a greater risk of obesity and obesity-related illnesses.

2a. Agricultural companies have failed to convince consumers to buy foods from crops whose genes have been modified.

2b. Agricultural companies have failed to convince consumers to buy genetically modified foods.

In the versions b of the two examples, relative clauses are replaced by modifier phrases, making the sentences more concise and readable. Here, two types of common modifier phrase patterns are used:

- *n.* + *v.*-ing/*v.*-ed/*adj.* + *n.* (as in *obesity-related illnesses*)
- *adv.* + *v.*-ing/*v.*-ed/*adj.* + *n.* (as in *genetically modified foods*)

Post-modifier

More complicated and longer modifier phases tend to appear after the modified nouns. Compare the two versions of the following sentences:

a. Instead, we focus on ***the demand shock*** [1]that is caused by ***a general loss of income*** [2]which affects consumers' spending patterns.

b. Instead, we focus on ***the demand shock*** [1]caused by ***a general loss of income*** [2]affecting consumers' spending patterns.

In the version b of the sentence, the relative clauses 1 and 2 are replaced with two modifier phrases. These modifier phrases take the following structure:

- *n.* + *v.*-ing/*v.*-ed

Task 2 Read the following titles and paraphrase the underlined part with relative clauses.

1) Plant inputs of carbon to metal-contaminated soil and effects on the soil microbial biomass
2) Temporary reduction in daily global CO_2 emissions during the virus forced confinement
3) Hijacking of host cellular components as proviral factors by plant-infecting viruses
4) Harnessing host ROS-generating machinery for the robust genome replication of a plant RNA virus

Task 3 Identify the modifier phrases in the following sentences and analyze their structures.

Model: The same trend appeared when an optimal N rate was applied with the ANFR increasing from 59.5% to 84% when following a late-planted soybean. (*adv.* + *v.*-ed + *n.*)

1) Researchers used the MicroResp™ method to assess pollution-induced community tolerance in the context of metal soil contamination.
2) Stock solutions (50 mM) of $PtCl_4$ and $NaAuCl_4$ were prepared in double distilled water (DD water) and diluted appropriately to reflect desired Pt and Au concentrations.
3) Evidently, rapeseed root exudates showed dose-dependent antimicrobial activity against the mycelial growth of P. parasitica var. nicotianae.
4) A similar trend was observed in Pt amended soil: high concentrations of Pt at 2,000 mg/kg increased PHOS activity by 90%, but inhibited the other enzymes.
5) The absorption of N during the vegetative growth stage contributes to rice development during reproductive and grain filling stages through translocation, resulting in an increase in grain yield and quality.
6) Community food data were purchased from InfoUSA and Nielsen and classified according to previously established protocol.
7) Those who live near fast-food restaurants and convenience stores drink more sugar

sweetened beverages (SSB).

8) There was a marginally significant association between HCP's advice on losing weight and frequency of salad consumption.

9) Rice cultivars used in this trial were LaKast, Diamond, and CL153, which were selected to represent well-adapted southern long-grain, pureline cultivars.

10) In contrast to these first two reactions of tissues to wounding (i.e. as physical objects or as biochemical solutions), the third response (i.e. as physiologically active tissues) involves the active participation of the tissue.

Task 4 Rewrite the following sentences by replacing the underlined part with modifier phrases.

Model: Platinum (Pt) and gold (Au) are <u>precious heavy metals that are widely used</u> in the jewellery, auto-mobile, chemical and medical industries.

→ Platinum (Pt) and gold (Au) are <u>widely used precious heavy metals</u> in the jewellery, auto-mobile, chemical and medical industries.

1) This pathogen is <u>a typical pathogen that is borne by soils</u> and that infects plants through the production of zoospores.

2) Soil samples were taken from each MG strip at the time of planting to quantify the soil-N credits accumulated at the start of <u>the season when rice was cropped</u> through the previous crop management system.

3) The pandemic has reminded us just how dependent we are on <u>a global food value chain which functions well</u> and how vulnerable we are to disruptions in this key sector.

4) Soil spiking with Pt largely enhanced more enzyme activities at higher concentrations than observed in <u>soils that are spiked with Au</u>.

5) Within each MG strip, each of the six sub-plots was randomly assigned an N rate from <u>the treatments which had been previously listed</u>.

6) Our analysis shows that the economic recession exerts downward pressure on prices, especially for <u>commodities to which a high value is added</u>, such as meat and dairy products.

7) The results of analyzing the gut microbiota community revealed that there were significant differences between NC mice and <u>mice that were treated with FL</u>.

8) Whether rotating rapeseed with tobacco increases the soil microbial activity and diversity, the populations of <u>organisms which are beneficial to plants</u>, and the antagonism

toward pathogens is interesting and should be evaluated further.

Task 5 Identify the modified words and their post-modifier phrases.

Model: The observed significant differences in consumption behaviors between those obtaining HCP's advice and those not getting similar advice were modest.

1) This paper analyzes the impacts on global agricultural markets of the demand shock caused by the COVID-19 pandemic and the first wave of lockdown measures imposed by governments to contain it.
2) It depends on a number of factors affecting supply and demand of all goods including agricultural commodities.
3) The International Food Policy Research Institute (IFPRI) estimates that the economic contraction in 2020 could increase the number of people living in extreme poverty by a staggering 20%, or 140 million people.
4) The meat sector is directly affected by lower demand for meat products resulting from lower incomes and by the substitution of cheaper (plant-based) sources of calories.
5) On a silt loam soil, the interaction of both planting date and MG of the previous soybean crop influenced the maximal grain yield achieved by the rice crop.
6) A rapid microtiter plate method is used to measure carbon dioxide evolved from carbon substrate amendments so as to determine the physiological profiles of soil microbial communities.
7) Table 2 shows associations between key independent variables and outcomes adjusted for age, gender, race/ethnicity, education, general health status, city of residence, BMI, and panel.
8) Most studies analyzing the influence of the community food environment have focused on fruit and vegetable consumption.
9) Amino acid sequence analysis using PrDOS (Protein Disorder prediction system), a natively disordered site-prediction web server, suggests that RCNMV MP, p27, and p88 have putative disordered domains.
10) The differences seen in soil N credits at rice emergence between previous soybean planting dates on a clay soil may have been due to the larger quantity of residue, a longer time span for mineralization to occur, and the warm winter climate experienced at SEREC.

Task 6 Complete the following sentences with the present or past participles (*v.*-ing or *v.*-ed) of the words given in the brackets.

Model: There will be secondary waves ___forcing___ (force) governments to impose new lockdown measures.

1) Soil samples were left overnight to equilibrate. Excess chloride (Cl) was removed from the soil ____________ (use) artificial rainwater.
2) Soil microorganisms are key components of soil and as such are an effective indicator of soil performance and "health" ____________ (cause) by pollution.
3) The contribution of this paper is a quantification of the impacts on the global agricultural markets and on the GHG emissions ____________ (result) from agricultural production.
4) The activity of key enzymes ____________ (involve) in the biogeochemical cycling of C, N, P and S in soil was assessed with seven different enzymes.
5) Heavy metals can be aggregated into minerals, or incorporated into basic microbial secondary metabolites ____________ (secrete) by microbes that interact with metal surfaces.
6) Brassica crops ____________ (rotate) with potatoes had higher microbial activity and diversity, which may help to suppress soil-borne diseases of potatoes.
7) Research on the factors ____________ (influence) the effectiveness of HCP weight loss advice have focused on the quality of counselling, but have not included patient-centered factors.
8) Leaching and aging decrease nickel toxicity to soil microbial processes in soils freshly ____________ (spike) with nickel chloride.
9) The results differed by soil texture, with the planting date and MG selection of the previous soybean crop more influential to the rice crop ____________ (produce) on a silt loam soil than a clay soil.
10) Educational interventions ____________ (target) medical students on weight bias and the causes of obesity combined with facilitated discussions with OW/OB patients have proven to be successful in creating short-term improvements in care.

2E Cohesion (1): *Conjunctions*

Readers interpret cohesive ties as meaningful and use them, together with context, to create coherence in what they are hearing or reading. They play a vital role in text organization and logical development of ideas, which is helpful for students when they read and write academic papers.

Conjunctions, as very useful and powerful cohesive ties in English, will be analyzed in detail in this unit.

1. Categories of conjunctions

There are four categories of conjunctions: coordinating conjunctions; subordinating conjunctions; correlating conjunctions; conjunctive adverbs or conjunctive adverbials.

a. Coordinating conjunctions

Coordinating conjunctions are used to link two clauses or phrases of equal value or equal status. They give equal value to the two elements that they coordinate and they must be placed between the two elements that they coordinate. There are seven coordinating conjunctions: FANBOYS —***for***, ***and***, ***nor***, ***but***, ***or***, ***yet***, ***so***.

Examples

1. The cuticle **and** epidermis of fruits **and** vegetables impose a resistance to the diffusion of gases (e.g. CO_2).
2. The lower clamp was removed **and** the head was padded with soft black foam to provide a seal **and** to avoid damage to the root surface during measurements.

3. When wound signals converge from two directions, there is a double induction, **but**, because of the distances and limitations of the tissues to respond, there is only a slightly increased physiological response.

b. Subordinating conjunctions

Subordinating conjunctions are used to link two clauses within a single sentence, when one clause is subordinate to the other.

In other words, the subordinate clause clarifies, expands or explains the meaning of the main clause. Common subordinating conjunctions include:

- ***as*, *because*, *since*, *so***
- ***although*, *though***
- ***after*, *before*, *until*, *while***
- ***if*, *unless*, *as long as*, *provided***
- ***whenever*, *whatever***
- ***in order that*, *so that***

Examples

1. **As** the slices ripened, the seeds exposed on their surfaces began to germinate.
2. When possible, these types of treatment should be avoided, **since** they add cost and complexity to an already complex system.
3. The effect on respiration is much more pronounced, **so that**, **as** the temperature increases, the concentration of O_2 decreases and the concentration of CO_2 increases within tissue.
4. **If** the gradient is insufficient to produce this flux, the internal CO_2 concentration will increase **until** a steep enough gradient is established.

c. Correlating conjunctions (also known as correlative conjunctions)

Correlating conjunctions are connectors that can correlate either words, phrases, or clauses (sentences). Common Correlating conjunctions includes: ***both ... and*, *not only ... but also*, *either ... or*, *whether ... or not*, *neither ... nor*, *the more ... the more ...*, *no sooner ... than***, etc.

Examples

1. This is **both** stupid **and** incomprehensible.
2. Temperature affects **not only** the rate of respiration, **but also** the rate of gas diffusion.
3. Host plant resistance as a management tactic is expressed **either** through resistance to thrips feeding **or** through resistance to thrips-transmitted viruses.

d. Conjunctive adverbs/adverbials

Common conjunctive adverbs/adverbials include: ***however***, ***moreover***, ***therefore***, ***thus***, ***besides***, ***meanwhile***, ***firstly***, ***basically***, ***for example***, ***in conclusion***, ***in contrast to***, ***in addition to***, ***as a result***, etc.

Conjunctive adverbs/adverbials are very similar to subordinating conjunctions. The biggest difference is that subordinating conjunctions are an indispensable part of a sentence, that is, without it the sentence is not complete, while a conjunctive adverb/adverbial is used when further information will be added. In addition, conjunctive adverbs/adverbials can frequently (but not always) be used in **a variety of positions** within the subordinate clause, whereas subordinating conjunctions **MUST** stand at the start of the subordinate clause.

Examples

1. This short article has only touched on a few of the many responses and interactions that occur during and after the processing of fresh-cut products. **For example**, unexpected responses may occur.
2. **In contrast to** these first two reactions of tissues to wounding (i.e., as physical objects, or as biochemical solutions), the third response involves the active participation of tissue.
3. This increased nutrient content allows WFT progeny to have greater survivorship and faster development than those developed on uninfected host plants. **Furthermore**, WFT reared on infected plants may have greater longevity and fecundity than those reared on uninfected plants.
4. **As a result**, the symptoms in the infected plant are phenocopy loss-of-function or reduced-expression mutants in the host gene.

2. Purpose of conjunctions

Conjunctions have different purposes, and eleven purposes of conjunctions are listed in Table 2.5.

Table 2.5 Purpose of Conjunctions

Purpose	Conjunctions
to coordinate/correlate	and, both ... and ..., either ... or ..., neither ... nor ..., not only ... but also ..., the more ... the more ...
to show contrast/concession	however, whereas, while, although, even though, in contrast, on the other hand, despite, in spite of, nevertheless, yet, if not, otherwise, conversely
to emphasize/highlight	certainly, indeed, specifically, in fact, to be specific, in other words, of course, naturally, definitely
to show a sequence / time relationship	when, while, firstly, secondly, thirdly, after, until
to show consequence	consequently, therefore, thus, for this reason, hence, accordingly, so
to add information / reinforce	additionally, also, besides, as well as, in addition, furthermore, moreover
to compare	similarly, in the same way, likewise, equally, by comparison, alternatively
to provide supporting information	to illustrate this, for example, for instance
for transition	regarding, with regard to, with respect to
to summarize	in conclusion, in summary, to conclude, to sum up
to show stance	basically, essentially, interestingly, surprisingly

Task 1 **Identify the *conjunctions* in the abstract and write them in the corresponding box.**

Title: *Functional and Sensory Properties of Phenolic Compounds from Unripe Grapes in Vegetable Food Prototypes*

(To be continued)

(Continued)

Abstract:

Unripe grapes (UGs) from thinning are an unexploited source of phenols useful as functional ingredient. However, phenols may negatively affect the sensory quality of food. Chemical and sensory properties of UG phenols in plant-based foods were not investigated before.

With this aim, an extract from UGs, obtained by a green extraction technique, was used to fortify three plant-based food models: carbohydrates/acidic pH/sweet — beetroot purée, proteins/neutral pH/sweet — pea purée and starch/neutral pH — potato purée.

Functional and sensory properties of phenol-enriched foods varied as a function of their composition and original taste. The amount of UG phenols recovered from potato purée was higher than that recovered from beetroot and pea purée, while the antioxidant activity detected in beetroot purée was higher than that in potato and pea purée. Significant variations of sourness, saltiness, bitterness and astringency were induced by UG phenols added to food models. Beetroot purée resulted was more appropriate to counteract the negative sensations induced by UG phenols.

Cohesive ties					
Grammatical cohesion			**Lexical cohesion**		
Conjunction	Substitution	Reference	Lexical reiteration: repetition, synonym, superordinate	Lexical chains	General Nouns

Task 2 Complete the sentences with the conjunctions given below.

while	not only ... but also ...	however	therefore	furthermore

1) Temperature affects ________ the rate of respiration, ________ the rate of gas diffusion. ________, the effect on respiration is much more pronounced, so that as the temperature increases, the concentration of O_2 decreases and the concentration of CO_2 increases within tissue.
2) Table 1a presents field and experimental information, ________Table 1b presents the cultivars included.
3) A better understanding of the wound healing process at lower humidity may contribute to efforts to extend the shelf life of root crops by improved handling and cultivar

selection. ________, the findings would be of general interest, as such conditions are typical of the environments under which plant tissue healing would normally occur.

4) The color of the inner tepals of wintersweet flower is an important criterion for variety identification. We, ________, focused on quantifying the change in the pigmentation of the inner tepals.

Task 3 Join the two simple sentences together with the coordinator *and*, *but*, or *or*.

1) Milk is a biological fluid produced by the mammary glands.
   Milk has received extensive attention due to its special nutrition and functions for humans.
2) The lactic acid bacteria are considered to be weakly proteolytic compared with many other groups of bacteria (Bacillus, Proteus, Pseudomonas, Coliforms) which may be present in the fresh meat and other ingredients.
   There is no direct evidence that these microorganisms contribute significantly to the flavor of fermented meat and it is doubtful whether their enzymes play any role at all in meat proteolysis.
3) After protein precipitation by zinc acetate, the mixture was centrifuged at 10,000 r/min and 4 °C for 10 min.
   The supernatant was used to determine the content of FAAs.
4) The experts can choose to record information.
   They can choose to neglect it.
5) Pickering and Gathercole (2002) used the Test for Children.
   They found an improvement in the working memory of the children they tested.

Task 4 Fill in the blanks with the conjunctions given in the list.

however	although	when	thus

Wounds in sweet potatoes heal most efficiently 1)________ the roots are exposed to temperatures of 28–30 °C and a relative humidity (RH) greater than 85% (Kushman & Wright, 1969). 2) ________ curing is practiced commercially in temperate areas, in the tropics it is often assumed that it takes place naturally (Collins & Walter,1985; Woolfe,

1992) and is not actively practiced. 3) ________, Jenkins (1982) reported that artificial curing under tropical conditions in Bangladesh did not reduce weight losses. 4) ________, the high levels of weight loss and very short shelf-life often seen in the tropics put into doubt whether wound healing takes place.

Task 5 Compare the two texts and identify the differences. Which one is easier to understand and why?

1) Due to the lack of data, we do not consider these supply-side disruptions to the agri-food sector in this paper. Instead, we focus on the demand shock caused by a general loss of income affecting consumers' spending patterns. The resulting lower demand obviously leads to a downward pressure on producer prices and production, but it is not clear, a priori, how large the effect will be in the different interdependent agricultural sectors. The meat sector, for example, is directly affected by lower demand for meat products resulting from lower incomes and by substitution towards cheaper (plant based) sources of calories. However, lower demand for grain and oilseeds also reduces feed costs, so the size of the net effect on production and prices is unclear.
2) Due to the lack of data, we do not consider these supply-side disruptions to the agri-food sector in this paper. We focus on the demand shock caused by a general loss of income affecting consumers' spending patterns. The resulting lower demand obviously leads to a downward pressure on producer prices and production. It is not clear, a priori, how large the effect will be in the different interdependent agricultural sectors. The meat sector is directly affected by lower demand for meat products resulting from lower incomes and by substitution towards cheaper (plant based) sources of calories. Lower demand for grain and oilseeds also reduces feed costs. The size of the net effect on production and prices is unclear.

Task 6 Select the correct expressions to complete sentences.

1) *In addition / For example* (supplementary Table S6), 8 amino acids were also identified, 7 of them being essential amino acids (W, L/I, T, V, K, R and F).
2) We are all aware that fresh cut produce is alive, with all its accompanying attributes, *so/ but* we often fail to consider what happens to the injured tissue as it responds to being wounded.
3) *Although/However* TRV: PDS-silenced leaves contain a higher proportion of white tissue

than PVX: PDS-silenced leaves, PDS mRNA levels are approximately the same.

4) There are a number of reasons why companies innovate. *Firstly/Specifically*, it is to maintain a competitive edge in the market. Secondly, it is to establish them to enter new markets.
5) *If/As* the slices ripened, the seeds exposed on their surfaces began to germinate.

2F Assignments: *Reading Comprehension and Vocabulary*

Part One Reading comprehension

Task 1 Read the following abstract and tell whether the statements that follow are true (T) or false (F).

gasdermins *n.* 焦孔素
permeabilize *v.* 使具渗透性
membrane *n.* 细胞膜
pyroptosis *n.* 细胞焦亡
proteolytic *adj.* 蛋白水解的；解蛋白的
autoinhibitory *adj.* 自抑制的
carboxy-terminal *adj.* 羧基端的
caspase *n.* 半胱天冬酶；切冬酶（胱冬肽酶）
pathogen *n.* 病原体，致病菌
Group A Streptococcus A组链球菌
secrete *v.* 分泌
protease *n.* 蛋白酶
virulence *n.* 毒性
cleavage *n.* 劈开，分裂
amino-terminal *adj.* 氨基末端的
keratinocytes *n.* 角质细胞
homologue *n.* 同系物
susceptible *adj.* 易得病的；易受影响的
hypervirulent *adj.* 具有高毒性的
exogenous *adj.* 外生的；外因的

Gasdermins (GSDMs) are a family of pore-forming effectors that **permeabilize** the cell **membrane** during the cell death program **pyroptosis**. GSDMs are activated by **proteolytic** removal of **autoinhibitory carboxy-terminal** domains, typically by **caspase** regulators. However, no activator is known for one member of this family, GSDMA. Here we show that the major human **pathogen Group A Streptococcus** (GAS) **secretes** a **protease virulence** factor, SpeB, that induces GSDMA-dependent pyroptosis. SpeB **cleavage** of GSDMA releases an active **amino-terminal** fragment that can insert into membranes to form lytic pores. GSDMA is primarily expressed in the skin 10, and **keratinocytes** infected with SpeB-expressing GAS die of GSDMA-dependent pyroptosis. Mice have three **homologues** of human GSDMA, and triple-knockout mice are more **susceptible** to invasive infection by a pandemic **hypervirulent** M1T1 clone of GAS. These results indicate that GSDMA is critical in the immune defence against invasive skin infections by GAS. Furthermore, they show that GSDMs can act independently of host regulators as direct sensors of **exogenous** proteases. As SpeB is essential for tissue invasion and survival within skin cells, these results suggest that GSDMA can act akin to a guard protein that directly detects concerning virulence activities of microorganisms that present a severe infectious threat.

______ 1) This abstract has all the BPMRC elements.

______ 2) This abstract starts with a purpose statement.

______ 3) This abstract is organized in specific-general pattern.

______ 4) GSDMA is a key word in the abstract.

______ 5) The findings of this research suggest that GSDMA is critical in the immune defence against invasive skin infections by GAS.

______ 6) Present tense is the only tense used in this abstract.

Task 2 Read the following abstract and answer the questions that follow.

① Plants depend upon beneficial interactions between roots and microbes for nutrient availability, growth promotion, and disease **suppression**. ② **High-throughput sequencing** approaches have provided recent insights into root **microbiomes**, but our current understanding is still limited compared to animal microbiomes. ③ Here we present a detailed characterization of the root-associated microbiomes of the crop plant rice by deep sequencing, using plants grown under controlled conditions as well as field cultivation at multiple sites. ④ The spatial resolution of the study distinguished three root-associated **compartments**, the **endosphere** (root interior), **rhizoplane** (root surface), and **rhizosphere** (soil close to the root surface), each of which was found to harbor a distinct microbiome. ⑤ Under controlled greenhouse conditions, microbiome composition varied with soil source and genotype. ⑥ In field conditions, geographical location and cultivation practice, namely organic vs. conventional, were factors contributing to microbiome variation. ⑦ Rice cultivation was a major source of global **methane** emissions, and **methanogenic archaea** could be detected in all spatial compartments of field-grown rice. ⑧ The

suppression *n.* 抑制

high-throughput sequencing 高通量测序

microbiomes *n.* 微生物群

compartment *n.* 隔层；分隔间

endosphere *n.* 根内部

rhizoplane *n.* 根面

rhizosphere *n.* 根际

methane *n.* 沼气；甲烷

methanogenic *adj.* 产甲烷的

archaea *n.* 古菌

co-abundance network 共丰度网络

microbial consortia 微生物总群

phyla *n.*（生物分类学的）门

depth and scale of this study were used to build **co-abundance networks** that revealed potential **microbial consortia**, some of which were involved in methane cycling. ⑨ Dynamic changes observed during microbiome acquisition, as well as steady-state compositions of spatial compartments, support a multistep model for root microbiome assembly from soil wherein the rhizoplane plays a selective gating role. ⑩ Similarities in the distribution of **phyla** in the root microbiomes of rice and other plants suggest that conclusions derived from this study might be generally applicable to land plants.

Questions

1) How many of these moves can you find in the abstract? Identify the sentences that correspond to each move in the table below and determine what tenses are used in each move.

Moves	Typical elements	Sentences	Tenses
Move 1	Background (B)		
Move 2	Purpose (P)		
Move 3	Methods (M)		
Move 4	Results (R)		
Move 5	Conclusion (C)		

2) There are no citations or references to previous research in this abstract. Is this common in your field?
3) Does the author of this abstract use the first-person pronouns (I or we)? What is the situation in your field?
4) What is the problem put forward at the beginning of the abstract?
5) What will happen to microbiome composition under controlled greenhouse conditions?
6) Identify the conjunctions used in this abstract. What is the purpose of their use?
7) Read the sentence 3 and write a sentence with the same structure "Here we ..., using ...".

8) Identify the head nouns in the phrases "the depth and scale of this study" , "conclusions derived from this study", and "a major source of global methane emissions".
9) Define the word "methane" with one of the sentence definition structures.

Task 3 Write an abstract for a research you have been involved in.

Part two Vocabulary

General academic vocabulary is used to refer to words that appear in texts across several different disciplines or to refer to lexical items that occur frequently and uniformly across a wide range of academic materials.

Task 1 Find out the general academic vocabulary in the following text and write them in the box.

Humic Substances (HS) from Leonardite and two different composts were used as biosurfactants to wash heavy metals (Cu, Pb, Zn, Cd, Cr) from a soil added with two metals concentrations and aged for 4 and 12 months. Composts were obtained by mixing manure with either 40% (CM-I) and 20% (CM-II) of straw as structuring material. For both aging periods and both metal concentrations, HS from CM-I removed more metals than from Leonardite, whereas the washing capacity of HS from CM-II was negligible. 13C-CPMAS NMR spectra of HS indicated that while aromatic moieties for CM-I and Leonardite were more abundant than CM-II, HS from CM-I was most abundant in carboxyl and phenolic carbons. Hence, HS from CM-I had a greater complexing capacity than from both Leonardite and CM-II and effectively displaced heavy metals from the soil during the washing treatment. Moreover, the amount of metals removed by solutions of ammonium acetate (AA) and diethylenetriaminepentaacetic acid (DTPA), was found invariably smaller than by HS from CM-I, thereby indicating that HS removed more than one metal species.

Type of vocabulary	Examples in the above text
General academic vocabulary	

Task 2 Match academic nouns 1)–10) to meanings a–j.

1) uptake	a. the quality or characteristic of something that makes it possible to approach, enter, or use it
2) accessibility	b. the division of something into smaller parts, or something that divides a space
3) partition	c. the process of taking food into the body through the mouth (as by eating)
4) amendment	d. be a signal for or a symptom of
5) allocation	e. a change to a law, either one that has already been passed or one that is still being discussed
6) indicate	f. an amount of money, space, etc. that is given to someone for a particular purpose
7) measurable	g. the act of making something different
8) validity	h. provide evidence for
9) modification	i. the quality of having legal force or effectiveness
10) manifest	j. capable of being measured

Task 3 Fill in the blanks with the academic words given in the box.

shown	available	sorb	amount	accumulation
Accordingly	bioaccessibility	essential	extensively	driven

As crop uptake of contaminants is 1)________ by their concentrations in soil pore water and plant transpiration, the 2) ________ of contaminants in soil pore water is the fraction 3) ________ for crop uptake. 4) ________, to reduce the uptake and 5) ________ of contaminants in crops, it is 6) ________ to decrease their concentrations in soil pore water. It is known that carbonaceous sorbents could 7) ________ many inorganic and organic contaminants from water, thus decreasing their bioavailability and/or 8) ________. Biochar, a carbonaceous material made from pyrolysis of biomass, has been 9) ________ studied as a soil amendment for sorption and immobilization of both inorganic and organic contaminants via surface adsorption, partition, and pore diffusion. It was 10) ________ that biochar amendment in soils could effectively decrease plant uptake of polychlorinated biphenyls, polycyclic aromatic hydrocarbons (PAHs), and pesticides.

Unit 3 Introduction

Main Contents	Learning Objectives
Subgenre: *Introduction*	★ Understanding the structure of an Introduction. ★ Identifying the structure of an Introduction.
Organization: *General-specific*	★ Grasping the development in a text from general to specific.
Rhetorical Function: *Problem and Solution*	★ Understanding problem and solution. ★ Recognizing components of situation, problem, solution, evaluation. ★ Learning to write problem and solution texts.
Register: *Tense in Citation*	★ Identifying tense in citations. ★ Learning to write sentences using different citation tenses.
Cohesion (2): *Reference*	★ Understanding cohesion. ★ Recognizing reference in paragraphs and texts. ★ Learning to write reference in paragraphs and texts.
Assignments: *Reading Comprehension and Vocabulary*	★ Reading: Understanding the Introduction better through in-depth reading. ★ Vocabulary: Learning collocations.

3A Subgenre: *Introduction*

An Introduction is the beginning and an essential part of an academic paper. The main purpose of an Introduction is to give background information on the topic of the research, briefly describe the purpose and significance of the research, review the previous studies in related fields, point out their research gap, and present the purpose and significance of the research.

As the Greek philosopher Plato remarked, "The beginning is half of the whole." An introduction is important because it needs to present the newest findings so as to grasp the attention of readers, especially editors for the purpose of publishing.

1. The structure of an Introduction

Although the requirements for the length and format of a paper vary from discipline to discipline, a typical introduction should provide: 1) the background of the study, 2) previous studies, 3) the study gap, 4) the purpose and significance of the study, and sometimes 5) the hypothesis of the study. The following table (Swales, 1990) shows the moves of an Introduction.

Table 3.1 Moves of an Introduction

Move 1 (Background)	Establishing a research territory	a) showing that the general research area is important, central, interesting, problematic or relevant in some way (optional) b) introducing and reviewing items of previous research in the area (obligatory)
Move 2 (Research Gap)	Establishing a niche	Indicating a gap in the previous research or extending previous knowledge in some way (obligatory)

(To be continued)

(Continued)

Move 3 (Present Research)	Occupying the niche	a) outlining purposes or stating the nature of the present research (obligatory) b) listing research questions or hypotheses (PISF) c) announcing principal findings (PISF) d) stating the value of the present research (PISF) e) indicating the structure of the RP (PISF)

Note: In ecology, a niche is a particular microenvironment where a particular organism can thrive. In our case, a niche is a context where a specific piece of research makes particularly good sense.

PISF = probable in some fields, but rare in others.

Example 1

Title of the paper: *Agroinfiltration: A Rapid and Reliable Method to Select Suitable Rose Cultivars for Blue Flower Production*

[A review about the previous studies] Flower color has always been of great interest from both floricultural and marketing point of views (Katsumoto et al., 2007). Cyanidin, delphinidin and pelargonidin are major anthocyanidins considered responsible for various flower colors. These compounds possess a different number of hydroxyl groups at their B-ring and as a result, the natural or artificial modification of the B-ring by hydroxylation could lead to changes in flower color. Flavonoid 3′-hydroxylase (F3′H) and flavonoid 3′, 5′ hydroxylase (F3′5′H), which are P450 enzymes, catalyze hydroxylation of the anthocyanin B-ring. The main substrates for B-ring hydroxylation are naringenin and dihydrokaempferol (DHK) (Forkmann and Martens, 2001). More specifically, the enzymes coded by the F3′H and F3′5′H genes convert DHK to dihydroquercetin and dihydromyricetin intermediates, respectively. Then, dihydroflavonol 4-reductase (DFR) catalyzes the reduction of the produced intermediates to cyanidin- and delphinidin-based anthocyanins, respectively (Holton and Cornish, 1995; Tanaka, 2006; Nishihara and Nakatsuka, 2010).

There are various naturally-occurring flower colors in the plant kingdom while some important colors have never been observed in certain plants, such as violet/blue colors in roses. This has been attributed to the absence of F3′5′H and consequently its product i.e. delphinidin-based anthocyanins (Tanaka,

2006). Conventional rose breeding programs used to modify flower colors have been found inefficient for developing blue color in roses because of the lack of required genes. Hence in the year 2007, genetic manipulation was employed for the first time to produce transgenic blue roses (Katsumoto et al., 2007). Since then, efforts aimed at improving flower colors by genetic engineering in plants have been intensified. However, the main setback faced has been the considerable time required for generating stable transgenic plants (Azadi et al., 2010; Kuriakose et al., 2012). To overcome this challenge and to facilitate the selection of a range of suitable genotypes for mainstream transformation, analyses of flavonoid composition using high-performance liquid chromatography (HPLC) were suggested (Katsumoto et al., 2007; Brugliera et al., 2013). **[Research gap]** However, this method is not sufficiently rapid and only provides a range of suitable genotypes for transformation rather than specifically determining them. **[The purpose of the present study]** Therefore, the present study was set to use agroinfiltration in evaluating flower color gene functions in order to identify suitable cultivars for generating blue rose flowers. More specifically, transient expression induced by agroinfiltration was performed by visual observations in a very short time while requiring minimum facilities.

Example 2

Title of the paper: *The Effect of Sheep and Cow Milk Supplementation of a Low Calcium Diet on the Distribution of Macro and Trace Minerals in the Organs of Weanling Rats*

[Showing that the general research area is important] Mineral deficiencies during youth and adolescence (up to the age of 25) impact the proper development of various anatomical systems (bones, soft organs etc.). In addition, chronic low-level mineral deficiencies result in short-term symptoms (lethargy, nausea, and weakness). As identified by Gharibzahedi and Jafari, the most commonly reported effects result from a lack of adequate iron (Fe), zinc (Zn), iodine (I), and selenium (Se). The diagnosis of deficiency is often related to short-term symptoms and ease of testing, which may not reflect the true rate in the population. For example, Fe deficiency has been identified at rates of up to 40% in preschool children, but methods for Fe deficiency testing vary, ranging from the visual hemoglobin color scale to the cyanmethemoglobin analytical colorimetric method. With respect to calcium (Ca), the occurrence of clinical deficiency is extremely rare, except in extreme dietary circumstances. The long-term impact of low level Ca deficiency during critical growth phases has an impact on growth rates, especially in the context of skeletal development, often leading to a restriction in bone growth.

[A review about the previous studies] The potential for mineral(s) deficiency leading to health complications in an individual can be related to a range of factors, including age, illness, allergy, genetic predisposition, and diet. Although it is often possible (and sometimes necessary in acute cases) for diet supplements to be used in the short to medium term, long-term diet supplementation is often seen as an ineffective treatment method. This is because mineral and/or nutritional dietary supplements are expensive, inefficient, can cause secondary health effects, and have the risk of over-supplementation. In comparison to long-term diet supplementation, the use of diet augmentation of food products is recommended.

[A review about the previous studies] Mineral deficiency is affected not only by total mineral intake but also by mineral absorption rates. Milk has been shown to have some positive effects on mineral absorption, independent of the high mineral concentration present. However, effects are likely to be different between milk from different species due to differences in composition. Specifically, cow milk (CM), buffalo milk, and goat milk may have beneficial effects on Ca absorption. Ca and phosphorus (P) are physiologically critical minerals, which are specifically required for growth and energy production. It is well established that dairy products generally play a key role in providing dietary Ca and P, typically through the consumption of CM, or CM derived products.

[A review about the previous studies] The trace and non-essential mineral profiles of sheep milk (SM) have been investigated to a limited extent. The concentration of selected toxic minerals in SM has been studied by Yabrir et al. and Ivanova et al. These studies identified that the concentration of these minerals is highly dependent on the location of milk production (i.e., environmental factors) and specific farming practices. We have previously provided some insight into the effects of New Zealand SM consumption in addition to a diet containing a balanced mineral profile. Despite SM having a higher mineral concentration and macronutrients (protein and fat content) compared to CM, its consumption did not have any effect on the development of rats fed a balanced diet. The livers of rats that consumed SM had a lower Fe content compared to CM fed animals. Higher concentrations of rubidium (Rb) and cesium (Cs) were present in the brain, kidney, liver, spleen, and serum of the rats fed SM compared to those fed CM. However, the absorption of all types of minerals (macro, trace, and non-essential minerals) occurs at different rates during dietary deficiency, dietary adequacy, and dietary excess. Therefore, this previous work gives insight into the effects of SM minerals in the context of a balanced diet with excess nutrients/minerals supplied by the milk. **[A review about the previous studies]** Previous studies have established the usefulness of using rat models fed Ca restricted diets to investigate mineral metabolism and bone structure. Rader et al. used weanling male rats to investigate the effects of a diet restricted in Ca (98% reduction) and in P (93%) on a range of serum biomarkers and reported that parathyroid hormone concentration was increased and vitamin D3 concentration was decreased during an 8-week feeding trial ($p < 0.05$).

[Research gap] There are no reports in the literature on the effect of SM consumption with a mineral-deficient diet. **[The purpose of the present study]** Therefore, this study was designed to investigate the effect of the consumption of SM or CM to a diet low in Ca and P on the macro, trace, and non-essential minerals distribution in the organs of weanling rats.

Example 3

Title of the paper: *Community Food Environment Moderates Association Between Health Care Provider's Advice on Losing Weight and Eating Behaviors*

[Showing that the general research area is important and problematic]

Overweight and obesity (OW/OB) continues to be a public health concern in adults, especially among low-income and minority populations. From 2011 to 2014, 68.5% of white adults were OW/OB, compared to 76.3% of non-Hispanic blacks and 78.4% of Hispanics (National Center for Health Statistics, 2016). Given the high prevalence of obesity, the U.S. Preventive Services Task Force recommends that health care providers (HCP) screen all adult patients for obesity and offer the appropriate treatment as needed (LeBlanc et al., 2011). **[A review about the previous studies]** Despite these national guidelines, only 47% of obese patients receive weight loss advice from their HCP (Loprinzi and Davis, 2016). Patients who receive weight loss advice from their HCP are more likely to eat less fat and calories (Bish et al., 2005; Loureiro and Nayga, 2006), and eat more salad and fruit (Lorts and Ohri-Vachaspati, 2016). **[Research gap]** However, not all who receive advice from their HCP change their eating behavior (Rodondi et al., 2006), suggesting that knowledge alone may not lead to behavior change. Research on the factors influencing the effectiveness of HCP weight loss advice has focused on the quality of counseling (Alexander et al., 2011), but has not included patient-centered factors.

[Extending previous knowledge to the present problem] The community food environment may be a potential moderator in the association between HCP weight loss advice and eating behavior. The food environment has been shown to be associated with eating behavior. Those who live near a supermarket or grocery store eat more fruits and vegetables (Dunn et al., 2015; Robinson et al., 2013), while those who live near fast food restaurants and convenience stores drink more sugar-sweetened beverages (SSB) (Laska et al., 2010). The confluence of exposure to environmental factors that promote healthy eating, alongside obtaining HCP advice, may result in healthier food consumption behaviors among OW/OB individuals. **[The purpose of the study]** Using data from a low-income, high-minority OW/OB sample, this study explores how the community food environment may moderate the relationship between receiving HCP's advice on losing weight and eating behavior, potentially explaining why some patients improve their eating behavior after receipt of weight loss advice while others do not. **[The hypotheses of the study]** There are two hypotheses. First, the association between receiving HCP's advice on losing weight and participants' frequency of fruit and vegetable consumption will be stronger among those who live closer to small grocery stores and supermarkets, compared to those who do not live close to these outlets. Second, the association between HCP's advice on losing weight and sugar-sweetened beverages and fast food consumption will be weaker among those who live closer to convenience stores and limited-service restaurants, compared to those who do not live near these food outlets.

Task 1 **Read the following Introductions and match features with sentence(s).**

Introduction 1

(Title of the paper: *Physiochemical Properties of Rice with Contrasting Resistant Starch Content*)

① Rice is one of the most important cereal crops in the world because the starchy endosperm provides energy in the human diet. ② It used to be generally believed that starch in the diet had to be totally digested in order to provide energy, but more is now known about resistant starch (RS), which is defined as starch and the products of starch degradation that cannot be absorbed in the small intestine of healthy individuals and might be fermented in the colon (Englyst et al., 1992). ③ With changes in consumer food preferences and lifestyles, resistant starch is attracting greater attention for its' health-promoting properties, i.e. a reduction in postprandial glucose and oxidative stress, preventing colonic diseases and reducing the risk of developing colorectal cancer through the production of short-chain fatty acids by fermentation (Zaman and Sarbini, 2016).

④ Resistant starch is divided into five types: RS1, RS2, RS3, RS4, and RS5. RS2 and RS3 are the main types found in raw and cooked rice respectively. ⑤ RS2 is composed of native starch granules from certain plants and is found in uncooked starch or starch that is not well gelatinised and hydrolysed slowly by α-amylases; RS3 represents retrograded or crystalline non-granular starch formed after cooking (Bird and Topping, 2008) and is widely found in food products today, providing the functional property of RS in the human digestive system. ⑥ Rice is consumed after cooking and consequently RS3 is the primary type found in cooked rice, ranging from 0.6% to 1.21% (Yang et al., 2006). ⑦ Owing to its health properties, an increase in the RS3 content in rice is desirable.

⑧ RS2 and RS3 are both naturally occurring starches, and their resistant ability can be stable during most physical food operations. ⑨ RS content is primarily determined by the starch source, its structure, i.e. granule size (GS), its types, and the fine structure of amylose and amylopectin such as chain length and branching (Lockyer and Nugent, 2017). ⑩ A higher amylose content of starch has been shown to be associated with increased resistance to enzymatic digestion (Lockyer and Nugent, 2017; Zhou et al., 2016). ⑪ Rice accessions with high RS starch generally show a different starch structure from their parental lines, such as higher percentages of granules with an irregular shape and smaller size (Yang et al., 2006), higher proportions of short (Shu et al., 2007) or long chain amylopectin (Tsuiki et al., 2016).

⑫ However, these studies have been conducted by different research groups using starch contained in plant materials that have been processed in different ways, e.g. many studies have been based on raw milled rice (RS2 type) while others have used cooked rice (RS3 type). ⑬ There are still gaps in scientific knowledge about the detailed relationships between RS2 and RS3, and their starch structure and composition. ⑭ The objective of the current study was to understand the physiochemical properties of rice accessions with differing amounts of RS2 and RS3, and to obtain physiochemical parameters that can be used for selecting rice cultivars that are high in RS3. ⑮ Due to the importance of resistant starch to human health, the research and classification of rice with different RS2 and RS3 contents will greatly benefit current food production and hence improve the consumers' lives.

1) Purpose of the study ________
2) Previous studies ________
3) General research area ________
4) Research gap ________
5) Hypothesis of the study ________

Introduction 2

(Title of the paper: *An Efficient Meshless Method for Bimaterial Interface Cracks in 2D Thin-layered Coating Structures*)

① Advanced structures containing thin-layered coatings are widely utilized nowadays in modern engineering devices and components. ② One of the important failure modes for such layered structures is the interfacial cracking or debonding of the coating from the substrate, due to the mismatch of mechanical properties between different materials.

③ One of the main difficulties for interface crack analysis arises from the extreme geometrical configuration of the thin-layered coating system, where the thickness of the thin-layered coatings is much smaller than that of the substrate. The well-established finite element method (FEM) is one of the dominant numerical methods for solving problems involving cracks. ④ However, within and near the thin-layered coatings, very fine meshes must be discretized to avoid high element distortion associated with the FEM. ⑤ During the past two decades, the boundary element method (BEM) has been viewed as an important alternative to the FEM for fracture mechanics analysis due to its boundary-only discretization and semi-analytical nature. ⑥ Yuuki and Cho used Hetenyi's fundamental solution for interface crack analysis in dissimilar materials. ⑦ Tan and Gao demonstrated the use of the quarter-point element method for solving crack problems. ⑧ Gu and Zhang presented a novel special crack-tip element method for interface crack analysis. ⑨ The BEM itself, however, also has some inherent shortcomings. ⑩ One numerical difficulty in applying the BEM is the accurate calculation of various nearly singular integrals when dealing with problems with crack-like and ultra-thin shapes.

⑪ Besides the FEM and BEM, in recent years, various meshless methods have been rapidly improved and can be nowadays considered as competing alternatives for fracture mechanics analysis. ⑫ Although impressive results have been obtained, no applications of the meshless methods to interface crack analysis in thin-layered coating structures have been reported so far, to the best of the authors' knowledge. ⑬ In the present paper, we model the fracture problems in thin coating/substrate structures with interface cracks by using the generalized finite difference method (GFDM), a recently developed meshless collocation method. ⑭ The method was introduced by Benito, Urena and Gavete and has been further improved by many other authors. ⑮ This method has already been used in crack analysis. ⑯ As a matter of fact, no element connectivity is needed and so, the method does not require a mesh to build the shape functions used to interpolate the displacement and stress fields. ⑰ We intend to demonstrate in this paper that the GFDM remains an accurate and efficient way for analyzing interface cracks in thin-layered coating/substrate composite materials under fairly general loading conditions.

⑱ The outline of the paper is as follows. ⑲ The definition of complex stress intensity factors (SIFs) of bimaterial interface cracks is briefly discussed in Section 2. ⑳ A displacement extrapolation method (DEM) is also introduced for the accurate evaluation of complex SIFs. ㉑ The equations of the GFDM are given in Section 3, where a multi-domain technique to deal with the non-homogeneity of the dissimilar materials is also discussed. ㉒ In Section 4, a benchmark numerical example is well studied to illustrate the accuracy and efficiency of the present method, where numerical results calculated by using the BEM are also given for the purpose of comparison. ㉓ Finally, some conclusions and remarks are provided in Section 5.

1) Purpose of the study ________
2) Previous studies ________
3) General research area ________
4) Research gap ________
5) Structure of the paper ________

Task 2 **Read the following Introduction 3, which is wrongly structured, and restructure it by writing the numbers in the correct order according to the moves.**

Introduction 3

① Against this background, the objective of this trial was to determine the extent to which pregnant sows can utilize the fibrous byproduct, oat hulls, and to measure its effects on performance.

② It is generally considered that the nutritive value of fibrous constituents for monogastric animals is low and that the digestibility of the diet decreases with increasing crude fibre. On the other hand, Cole et al. concluded that the provision of fibre may have some nutritional significance for the pig, especially for the sow having a fully developed alimentary tract containing a well-developed population of microorganisms.

③ Although extensively studied with growing-finishing pigs, the utilization of fibrous feedstuffs in sow diets has received relatively little attention.

④ There has been considerable interest in the use of low-cost fibrous ingredients in pig diets as alternatives to cereals.

⑤ In addition to their potential as sources of energy, fibrous constituents may also be of value in allowing increased feed intake by reducing nutrient density in sow diets. Practical feeding systems for sows in pregnancy generally involve the use of a diet of moderate energy density in the region of 12.5 MJ/kg. The amounts given are invariably well below the ad libitum intake, in order to prevent excessive body mass gains and fat deposition and to maximize appetite in subsequent lactation to ensure an adequate milk yield. The resulting low degree of satiety in the sow during pregnancy may contribute to behavioural problems.

Correct order: ________________

2. Sentence structures frequently used in an Introduction

In each move of an Introduction, there are some fixed expressions, which are helpful in writing an Introduction or identifying the elements of an Introduction.

Table 3.2 Sentence Structures Frequently Used in an Introduction

Moves	Sentence structures
Move 1 (Background)	... has received/drawn/attracted considerable/increasing attention. There has been an increasing / a growing interest in ... Researchers have shown an increased interest in ... ... has become a favorite topic for ... In the new global economy, it has become a central issue ... ... is becoming increasingly important/an ever-growing problem ... There has been an explosive growth in ... There has been / is an increasing amount of research ... In recent years, there has been a great deal of interest in ... In most studies of ..., ... has been emphasized with attention given to ... Several researchers have theoretically investigated ... In the last decade, there has been a growth in ... The recent development of ... has heightened the need for/has led to ... It has been increasingly recognized as a serious, worldwide public health concern in the field of ... ... has been studied extensively in recent years. Many investigators have recently turned to ... Many recent studies have focused on ... In studies, researchers have concluded that ... Previous studies have demonstrated/shown that ... In the past decades, a number of researchers/investigators have reported/examined ... There have been a number of studies involving ...
Move 2 (Research gap)	However, this is inconsistent with this argument ... However, the previously mentioned methods suffer from some limitations/weaknesses/disadvantages ... However, few writers have been able to ... However, these results were limited to ... However, previous studies/researchers have not dealt with ... However, ... remains unclear/unknown ... The existing literature fails to resolve/explain ... Little attention has been paid to ... There are relatively few studies on ...

(To be continued)

(Continued)

Moves	Sentence structures
Move 3 (Present research)	The purpose/aim/objective of this paper/investigation/study was to determine/analyze/evaluate/establish ... The aim/purpose of this paper/study/investigation is to apply/provide/review/give detailed information on ... In order to / To address / To tackle this problem of ..., we have developed / we propose ... In this paper, ... is discussed/studied/presented/investigated/described. The present study therefore focuses on ... This paper reports on ... The present work mainly deals with ...

Task 3 **Reread the three Introductions (1–3) above and find out the expressions that helped you to identify the elements (BPRO).**

Introductions	General research area	Previous studies	Research gap	Purposes
Introduction 1				
Introduction 2				
Introduction 3				

3B Organization: *General-specific (GS)*

General-specific (GS) is a common skill in academic writing because it is comparatively simple, and is often used in an Introduction for a longer piece of writing. Its structure involves general-to-specific movement which is similar to the shape of a glass or cup. When we need to provide an answer to an examination question, give an opening paragraph for an assignment or a background paragraph for an analysis or discussion, we usually produce a GS text.

Example 1

① Indole-3-acetic acid (IAA) is the most common auxin-class phytohormone and plays vital roles in plant growth and plant development processes, <u>such as</u> cell division, cell expansion, cell differentiation and fruit development. IAA homeostasis is important for maintaining the hormonal balance at an optimum level suitable for normal plant growth and development. ② <u>However</u>, high levels of IAA can exert an inhibitory effect on plant physiological processes. Several studies have reported that IAA at high concentrations inhibits seed germination and plant growth. The inhibition scenario is caused by ethylene production due to aminocyclopropane-1-carboxylic acid synthase (ACC synthase) activity stimulated by high levels of auxin accumulation, which results in an ethylene burst that causes plant growth abnormalities and senescence. ③ <u>A more important factor</u> implicated in growth inhibition and the actual phytotoxic response to auxins is the overproduction of abscisic acid (ABA). This is why IAA is naturally synthesized at low levels in plants or derived from plant-associated microbes.

This GS text starts with a definition of indole-3-acetic acid (IAA) and its functions. Following the statement, specific information is provided using logical connectors, for instance, *such as*, *however*, *a more important factor*. Skillfully, the author presents more specific information on IAA with the help of conjunctions.

Example 2

① Food insecurity is evidently an important problem in middle- and low-income countries with important implications for public health. Recently, US workers have developed the concept of food insecurity as it applies to the populations of high-income countries. ② Food insecurity has been defined as the "limited or uncertain availability of nutritionally adequate safe foods, or limited or uncertain ability to acquire foods in socially acceptable ways". ③ Food insecurity is socially defined and includes problems with the quantity and quality of the food available, uncertainty about the supply of food, and experiences of going hungry. ④ Experiences of food insecurity include running out of food, running out of money to buy food, skipping meals, experiencing hunger and being unable to buy food, or buying cheaper foods because of financial constraints. ⑤ Estimates suggest that up to 12 per cent of US households may experience some degree of food insecurity. ⑥ Food insecurity has been associated with unfavourable food choices, and it has been suggested that food insecurity may predispose to the development of obesity.

The above GS text begins with a generalization statement. From the beginning to the end, we can clearly see how the author organize the information from general to specific:

Sentence ① introduces the topic of food insecurity.

Sentences ②&③ give the definitions of food insecurity.

Sentence ④ introduces the experiences of food insecurity.

Sentence ⑤ gives the specific experiences of food insecurity in the US.

Sentence ⑥ points out the specific reason and the possible specific outcome of food insecurity.

Example 3

① The information given by the three-line graph and the table is about the worldwide use of water and how water was consumed in two different nations in 2000.

② According to the line graph, all of the three sectors saw an upward trend during the 10 decades. ③ Exactly, agricultural water use was consistently higher than others, rising from about 500 m^3 in 1900 to 3,000 m^3 in 2000. Before 1950, industrial water use was similar to domestic one, remaining stable at just under 50 m^3. ④ However, during the next five decades, industrial use increased more sharply than domestic use, reaching about 1,000 m^3 in 2000.

⑤ As can be clearly seen from the table, Brazil, with 176 million people, had 26,500 m^2 of irrigated land, which was 265 times as much as the Democratic Republic of Congo. ⑥ Water consumption per person hit 359 m^3. ⑦ In contrast, that in Congo was only 8 m^3, although there were only 5.2 million people in that nation.

In the text, we can see the information included in each of the sentences:

Sentence ① summarizes the contents of the entire table, and points out the object, place and time.

Sentences ②–④ present overall trends: 10-year increase in both industrial, agricultural, and domestic water use; then describe the growth in agricultural water use, which has been at the highest level; finally compare the growth trends in industrial and domestic water use.

Sentences ⑤–⑦ describe the population, irrigated land area, and per capita water use in Brazil and Congo.

Task 1 Read the following text and answer the questions.

① Salinity is a major adverse environmental factor for plant growth, limiting the utilization of about 830 million ha of agricultural land globally, of which about 80 million ha of irrigated and dryland agriculture are seriously affected worldwide. ② This is one of the most important land resource problems for global food production, especially with respect to the rapidly increasing demands due to global population growth. ③ The ultimate solution to such problems is probably the development of salt-tolerant plants based on a comprehensive understanding of the salinity effects on plant biochemistry and plant adaptation mechanisms at the systems level.

1) What type of information is included in each sentence in this text?

____

2) How is this passage organized?

____

Task 2 The following are the sentences of a GS passage. Read the sentences and put them back in the correct order.

① Newer generation HTPs release nicotine and other volatile compounds by heating the tobacco rod at temperatures not exceeding 350 ℃.

② However, earlier generation HTPs that were marketed as less harmful alternatives to combustible tobacco still produced high levels of carcinogens and more CO than conventional cigarettes.

③ HTPs generally consist of a holder (a battery and an electronically controlled heating element) and a rod (containing processed tobacco, glycerine, and other additives).

④ This is in contrast to electronic cigarettes (ECs), which heat a tobacco-free nicotine solution and do not elevate CO levels.

⑤ At these temperatures, tobacco combustion is unlikely and as a consequence far fewer chemical toxicants (including CO) are formed.

⑥ Heated tobacco products (HTPs) — the newest addition to the smoking harm reduction approaches — are an emerging class of nicotine delivery devices that do not burn tobacco.

Correct order: ________________

Task 3 Read the following GS text and finish the exercises.

Salinity has detrimental effects on almost all aspects of plants, including seed germination, plant development and growth. These effects are related to the activation of salinity-induced molecular networks involved in stress perception, signal transduction, regulation of stress-related genes, protein expression and subsequently metabolism. For example, salt stress alters many kinase-based signal transduction pathways of plant cells, such as mitogen-activated and calcium-dependent protein kinases, glycogen synthase kinase and histidine kinase signaling. It is also known that salinity disturbs the ion and osmotic homeostasis, induces oxidative stress, affects plant hormone biosynthesis and alters metabolisms such as photosynthesis.

1) Which sentence is the general statement?

____

2) How many examples are mentioned? And what are these examples?

____

3C Rhetorical Function: *Problem and Solution*

Based on what we've learned in Unit 1, rhetorical functions can be classified into three main macro-functions: to describe, to explain, or to persuade. Within each of these broad categories, more specific functions which relate to them are listed. Problem and solution is one of the specific rhetorical functions that aim to persuade.

1. Structure of problem and solution text

Problem-solution texts consider the problems of a particular situation, and give solutions to those problems. They are in some ways similar to cause and effect texts, especially in terms of structure. Problem-solution texts are actually a subtype of another type of essay, which has these four components: **situation**, **problem**, **solution**, **evaluation**.

The "situation" may be included in the text prompt, in which case it will not be needed in the main body. If it is needed, it can often be included in the introduction, especially for short essays. The "evaluation" may be included as part of the conclusion, or omitted altogether, especially for short essays.

There are two main ways to structure a problem-solution essay: block and chain, which are similar to the ways to structure cause and effect essays. In the block structure, all the problems are listed first, followed by all the solutions. In the chain structure, each problem is immediately followed by the solution to that problem. Both types of structure have their merits. The former is generally clearer, especially for shorter essays, while the latter ensures that any solutions you present relate directly to the problems you have given.

The two types of structure, block and chain, as shown in the diagram below, are for a short essay that includes "situation" in the introduction and "evaluation" in the conclusion. A longer essay, say one of around 1,000 words with citations, would probably have these two sections as separate paragraphs in the main body.

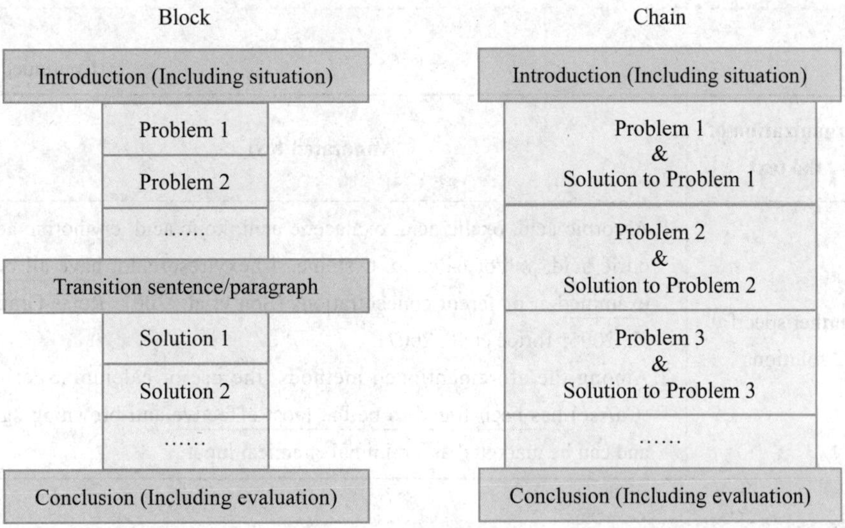

Example 1

Organization of the text	Annotated text
Background	Apples are one of the most widely consumed fruits (FAOSTAT, 2011) and are a good source of phytochemicals (Boyer and Liu, 2004). Epidemiological studies have linked the consumption of apples with reduced risk of some cancers, cardiovascular disease, asthma, and diabetes (Boyer and Liu, 2004)
Problem	However, fresh-cut processing results in major tissue disruption of surface cells and injury stress of underlying tissues (Toivonen, 2004). The main problem with fresh-cut apples is oxidation caused by polyphenol oxidase (PPO) that exists in particularly high amounts in apples (Whitaker, 1972). The resulting browning makes the product unsuitable to the consumer.
Proposed general solution to the problem	One way to increase fresh fruit consumption is to process fruit into fresh-cut products to be sold as convenient single servings.
Specific solution	Ranges of treatments have been applied to extend the shelf life of fresh-cut apples, mainly dipping in solutions of a wide range of anti-browning agents.

(To be continued)

(Continued)

Organization of the text	Annotated text
Further specific solutions	1. Ascorbic acid, oxalic acid, oxalacetic acid, kojic acid, erythorbic acid, citric acid, and/or calcium, cysteine, 4-hexylresorcinol have all been examined at different concentrations (Son et al., 2001; Rojas-Grau et al., 2006; Tortoe et al., 2007). 2. Among the aforementioned methods, the use of calcium ascorbate (CaAsc) has been found to be the most effective anti-browning agent and can be marketed as a minimal chemical input.

The above problem-solution text starts with the background of the problem and followed by the problem. The writer provides solutions from general to specific ones. From the information on the right, we can clearly see how the writer persuades the readers by supporting several solutions to the problem with fresh-cut apples.

Example 2

[Background] In the UK, people in low-income households have low intakes of fruit and vegetables and low micronutrient intakes. For these households, sources of lower priced, healthy food may be relatively inaccessible. Morris et al. recently estimated that the cost of achieving recommended dietary intakes would account for between 20 and 24 per cent of the minimum income required for healthy living. Their study estimated that the minimum income required for healthy living was probably higher than the UK national minimum wage.

[Problem] Food insecurity is evidently an important problem in middle- and low-income countries with important implications for public health. Recently, US workers have developed the concept of food insecurity as it applies to the populations of high-income countries. [Definition and examples] Food insecurity has been defined as the "limited or uncertain availability of nutritionally adequate safe foods, or limited or uncertain ability to acquire foods in socially acceptable ways". Food insecurity is socially defined and includes problems with the quantity and quality of the food available, uncertainty about the supply of food, and experiences of going hungry. Experiences of food insecurity include running out of food, running out of money to buy food, skipping meals, experiencing hunger and being unable to buy food, or buying cheaper foods because of financial constraints. [General solution] Estimates suggest that up to 12 per cent of US households may experience some degree of food insecurity. Food insecurity has been associated with unfavourable food choices, and it has been suggested that food insecurity may predispose to the development of obesity.

The above problem-solution text starts with the background of the problem and followed by the problem. The writer provides general solutions but not specific ones. We can clearly see how the writer persuade the readers by supporting general solution to the problem. The specific solutions could be added.

Task 1 **Read the following text and answer the questions that follow.**

① Traditionally, raw swine manure has been used to provide nutrients for plant growth and to improve soil conditions. ② However, the increase in concentrated animal feeding operations (CAFOs) results in high levels of nutrients in the proximal crop and pasture lands due to the production of more manure than required to meet the local plant nutrient demand (Ro et al., 2014). ③ Soil runoff and leaching of land-applied nutrients can enrich surface and groundwater with nitrogen and phosphorus compounds leading to eutrophication and hypoxia (Rabotyagov et al., 2014). ④ In addition, over application of animal manure can spread pathogens, release hormones and other pharmaceutically active compounds, and emit ammonia, greenhouse gases, and odorous compounds (Stone et al., 1998; DeSutter and Ham, 2005; Gerba and Smith, 2005). ⑤ Recently, the potential of thermochemical conversion of animal manures blended with other agricultural residuals to produce energy and/or biochar has been reviewed (Ro et al., 2010, 2014). ⑥ Although thermal pyrolysis of raw manure alone does not produce enough energy to support the conversion process, blending with other feedstocks with high energy density such as dried biomass or agricultural plastics can increase energy output enough for both biochar and power production (Ro et al., 2010, 2014). ⑦ Conversion of CAFO's surplus manures into biochar via pyrolysis is an alternative for manure management that may offer multitudes of environmental benefits (He et al., 2000). ⑧ Pyrolyzing manure destroys pathogens and substantially reduces odor and the volume of manure for easy handling, storage, and transportation (Pham et al., 2013). ⑨ In addition, manure-based biochar may be used as a soil amendment to improve soil quality as other plant-based biochar (Uzoma et al., 2011).

1) Which sentence depicts the background?
2) Which sentence depicts the main problem?
3) Which sentence depicts the solutions to the main problem?

Task 2 Read the following text and fill in the form with corresponding sentence numbers.

① Tobacco is the single most preventable cause of death in the world today. ② This year, tobacco will kill more than five million people — more than tuberculosis, HIV/AIDs and malaria combined. ③ By 2030, the death toll will exceed eight million a year. ④ Unless urgent action is taken, tobacco could kill one billion people during this century.

⑤ Most of tobacco's damage to health does not become evident until years or even decades after the onset of use. ⑥ So, while tobacco use is rising globally, the epidemic of tobacco-related disease and death has just begun. ⑦ But we can change the future. ⑧ The tobacco epidemic is devastating — but preventable. ⑨ The fight against tobacco must be engaged forcefully and quickly — with no less urgency than battles against life-threatening infectious diseases. ⑩ We can halt the tobacco epidemic and move towards a tobacco-free world — but we must act now.

⑪ The WHO Framework Convention on Tobacco Control, a multilateral treaty with more than 150 Parties, was the first step in the global fight against the tobacco epidemic. ⑫ This treaty presents a blueprint for countries to reduce both the supply of and the demand for tobacco. ⑬ The WHO Framework Convention establishes that international law has a vital role in preventing disease and promoting health.

Function	Sentences
Background to the problem	
Problem	
Cause of the problem	

(To be continued)

(Continued)

Function	Sentences
Effect of the problem	
Proposed general solution to the problem	
Specific solutions	
Evaluation of the solutions	

Task 3 The following are the sentences of a problem-solution text. Read the sentences and put them back in the correct order.

① As a result, the Loess Plateau, which was once covered with grass and forests, was turned into a barren land that supported fewer plants.

② In the early years, the Chinese cut down a lot of trees, at the source of the Yellow River, which led to the disappearance of large forests and to terrible floods.

③ Because many chemical factories discharged poisonous gas without filtering it, the cities were covered with so much poisonous gas that people were killed by the air they breathed in.

④ Only by changing the way we treat the environment can we get along well with it.

⑤ Third, enterprises should pay special attention to the effect they have on the environment and work out solutions to the problems.

⑥ Another example was in several developed countries.

⑦ Measures should be taken to protect the environment on which we depend.

⑧ Second, voices should be raised to inform the public of the importance of protecting the environment.

⑨ First, governments should prohibit the destruction of vegetation, rivers and lakes, oceans, and the atmosphere.

Correct order: ____________________

2. Sentence structures frequently used in a problem-solution text

There are some sentence structures frequently used in a problem-solution text, which may help you identify or write problem-solution texts.

Table 3.3 Sentence Structures Frequently Used in a Problem-Solution Text

Problems	Nowadays, ... has become a problem we have to face. Nowadays, more and more people are concerned about the problem of ... Recently, the rise in the phenomenon of ... has drawn/aroused/captured worldwide attention. Recently, the issue/problem/question of ... has been brought into focus / brought to public attention / posed among public attention. In recent years, many cities/nations/people have been faced/confronted with the serious problem of ... With the increasingly rapid economic globalization and urbanization, the problem of ... is being brought to our attention. Nowadays, an increasing number of people come to realize / be aware of the importance of ... With the rapid/amazing/fantastic development/growth/improvement of the economy, great changes have taken place in ... It is undeniable / There is no denying the fact that ... has become the utmost concern among people nowadays.
Reasons	There are many / several / a number of / a variety of reasons for this. First, ... Second, ... Last ... The reasons can be listed as follows. For one thing, ... For another ... What has possibly contributed to this problem? In the first place ... In the second place ... Perhaps the main / primary / most important reason is ... ... is also/partly/solely/chiefly responsible for the problem. A number of factors could account for / lead to / contribute to / result in / be conducive to ... This problem may result from / be attributable to / be largely due to / stem from a combination of factors.
Solutions	The first key factor to solving this problem is ... Another key factor is ... People have found many solutions/approaches to ... We can figure out many ways to alleviate/handle/address the current crisis. Faced with ..., we should take a number of effective measures to deal with the situation. It is high time that ... Here are a few examples of some of the measures that could be taken immediately. Obviously, if we ignore / are blind to this problem, the chances are that / it is most likely that / there is every chance that ... will be put in jeopardy. We need to take a fresh look at this matter/situation from a new perspective / a broader standpoint. Otherwise, we won' t come anywhere near solving it. It is essential/indispensable that proper actions / effective measures should be taken to reverse this situation / trend. It is hoped/suggested that considerable/great efforts should be directed to / focused on ...

3D Register: *Tense in Citation*

Citations are used to create a research space for the citing author by citing the viewpoints, materials and methods in other people's works and papers as the basis for his or her own arguments. The citing author can point out what has not been done through citations, thus highlighting the limitations of the previous study and preparing a space for new research.

Three methods are often used in citations: quoting, paraphrasing and summarizing (Cai, 2020:62).

Table 3.4 Methods of citations

Methods	Purposes	Examples
Quoting	Copying short, important sentences or passages from the original text word for word. (e.g. definition, ideas, explanations, rationales)	Writing has been described as the "ultimate juggling act" (Stainthorp, 2015).
Paraphrasing	Rewriting short extracts where there is no particular reason to quote the original words.	Feak (2011) encouraged students to engage with multidisciplinary studies.
Summarizing	Summarizing the main ideas/findings of an author or the essence of their arguments.	The research of Gardner (2012) demonstrates that the essay retains a strong position in student writing.

Tense choice in reviewing previous research is subtle and somewhat flexible. Several studies have shown that at least two-thirds of all quoted statements fall into one of the following three major patterns:

1) Past — researcher activity as agent

e.g. Jones (1987) investigated the causes of illiteracy.

The causes of illiteracy were investigated by Jones (1987).

2) Present Perfect — researcher activity not as agent

e.g. The causes of illiteracy have been widely investigated (Jones, 1987; Ferrara, 1990; Hyon, 1994).

There have been several investigations into the causes of illiteracy (Jones, 1987; Ferrara, 1990; Hyon, 1994).

3) Present — no reference to researcher activity

e.g. The causes of illiteracy are complex (Jones, 1987; Ferrara, 1990; Hyon, 1994).

The first and second are most common in the humanities and least common in science, engineering, and medical research. However, all three tend to occur in many extensive literature reviews, as they add variety to the text.

In summary, there are three patterns:

Pattern 1: The past tense

The simple past tense is often used in literature reviews or Introductions to refer to the findings of a previous study. In pattern 1, attention is given to what previous researchers have done.

Examples

1. Descriptions of the process of wound healing in sweet potato date from the 1920s when Weimer and Harter (1921) described how moisture and temperature affected the wound periderm formation and the efficiency of the wound cork in preventing infection.
2. Norman, Guindo, Wells, and Wilson (1992) found 79% of the fertilizer applied pre-flood was taken up by panicle differentiation, significantly higher than other cereal crops.

Pattern 2: The present perfect tense

The present perfect tense is preferred in the following situations:

a. The research referred to is recent, or is more indicative of the current state of knowledge than the research referred to earlier.

b. There is no specific date or time you can refer to at the beginning of the literature review. But after that, you can use the simple past to give specific examples in that particular area.

c. It is used when summarizing an area of research in the literature review.

d. It is used when referring to someone's views, position or argument, especially when comparing positions in a debate, and when that debate is still going on in current journals.

Examples

1. Although many studies have examined plant traits that correlate with competitive ability (Grime, 1974; Westoby, 1998; Keddy et al., 2002), it is unclear whether the traits evolved as a result of selective pressure from competition.
2. Life history characterizes factors that help organisms to survive and establish populations (Varley and Gradwell, 1970; Morse and Hoddle, 2006), but few studies on the life history of this pest have been published (Murai, 2001; Zhang, 2013).

Pattern 3: The present tense

The present tense is often used when referring to a work written a long time ago, but then this indicates that the work is so influential on current thinking. In pattern 3, the focus is on what has been found.

Examples

1. Those who live near a supermarket or grocery store eat more fruits and vegetables (Dunn et al., 2015; Robinson et al., 2013), while those who live near fast food restaurants and convenience stores drink more sugar-sweetened beverages (SSB) (Laska et al., 2010).
2. Chin (2013) claims that nutrient deficiency and tastelessness are two aspects of white rice which make it a less than ideal food for over 3 billion Asians whose calories are heavily dependent on it.

Task 1 Complete the following paragraph with the verbs in correct tenses.

In 1965, Vasil et al. (1965) 1) ________ (report) that a mature plant can 2) ________ (obtain) from a single vegetative cell of the tobacco for the first time. Furthermore, Horsch et al. (1985) 3) ________ (explore) the transformation in tobacco plants via Agrobacterium tumefaciens vectors. The use of genetic transformation for gene over expression in transgenic tobacco 4) ________ (increase) markedly over the years. Foreign DNA integration into plant cells in genetic engineering techniques 5) ________ (provide) a new approach to studying the function and structure of plant genes for cultivar improvement (Fraley, Roges, Horsch et al., 1986). Tobacco transformation efficiency of plants mostly 6) ________ (rely) on the physiological condition of tobacco cells, a tumefaciens strain and plasmid DNA. A general transformation method, through tobacco leaf disks, 7) ________ (use) by several researchers successfully (Gallois and Marinho, 1995).

Task 2 Complete the following paragraph with the verbs in correct tenses.

Honeybees 1) ________ (be) crucial for global agroecosystems, which 2) ________ (contribute) over $200 billion in pollination services for approximately 2/3 of crop species and most wild flowering plants (Connolly, 2013; Palmer et al., 2013; Rundlöf et al., 2015). These contributions of honeybees in agricultural products 3) ________ (increase) steadily, especially for those crops with highly economic values and nutritional content (Lautenbach et al., 2012; Stanley et al., 2015). Nonetheless, numerous studies 4) ________ (indicate) a negative correlation between honeybee health and neonicotinoid pesticides exposure, illustrating dramatic reductions in honeybee foraging (Feltham et al., 2014; Henry et al., 2012; Kessler et al., 2015), reproductive performance (Rundlöf et al., 2015; Whitehorn et al., 2012), and colony survival (Goulson, 2015; Stanley et al., 2015). Generally, much of the crisis in honeybee populations 5) ________ (attribute) to neonicotinoid pesticides, the most widely used chemical class of insecticides, which 6) ________ (account) for 80% of the global insecticidal seed treatment market share in over 120 registered countries (Jeschke et al., 2010). The main measures adopted to minimize the risk of neonicotinoids 7) ________ (be) the restriction in use of these pesticides in agroecosystems, especially in the European Union and in the United States (Gewin, 2008).

3E Cohesion (2): *Reference*

Cohesion is the act or state of linking something (in the text) with what has gone before. These links or ties are not only just within the sentence level but also beyond the sentence level. Cohesion is very meaningful and useful for readers to understand the context by creating coherence in what they are hearing or reading.

Reference is an effective way of creating coherence in academic writing. Making reference becomes possible by using demonstrative pronouns, demonstrative verbs, reflexive pronouns or relational pronouns to replace or indicate the words already mentioned. Reference is frequently used in research papers because it can help to achieve simplicity and avoid repetition in writing.

Nominal reference is used to refer to words and content that have already appeared in a context or sentence with alternative nouns or pronouns in order to avoid repetition and to connect the context. It is an important principle of English speaking and writing.

Personal pronoun, demonstrative pronoun and comparative pronoun can be considered as methods of nominal reference. The methods are specifically listed in Table 3.5.

Table 3.5 Methods of Nominal Reference

Methods of reference	Words
Personal pronoun	he, she, they, it, him, her, them, his, their, its, himself, herself, themselves, itself, who, whom, whose
Demonstrative pronoun	that, this, those, these
Comparative pronoun	the former, the latter

Examples

1. Rapeseed roots can attract **zoospores** into the rhizosphere and then secrete a series of antimicrobial substances to kill **them**.

Zoospores ⟵ **them**

2. **Antimicrobial activities** were further selected to determine **their** concentration in the rapeseed root exudates.

   Antimicrobial activities ⟵ **their**

3. **Plant roots** can also secrete a number of substances to protect **themselves** against pathogen and non-host pathogen infection.

   Plant roots ⟵ **themselves**

4. It is not very easy to fully understand the relationship between the scenic spot mode and **the whole region tourism mode**, in that **the latter** was supposed to revolve around the unique orientation of scenic spots.

   the whole region tourism mode ⟵ **the latter**

Task 1 Complete the following sentences with the reference words listed in Table 3.5.

1) The penaeid prawn Melicertus kerathurus (Forsköl, 1775) is a highly valued decapod crustacean that is fished throughout the Mediterranean and the West Atlantic Coasts, from Angola to Britain. ________ species can command a high market price and the correct management of the stock is important for the sustainability of the local fisheries and the communities who rely on them.
2) The above data demonstrated that rapeseed can secrete many antimicrobial compounds through ________ root exudates to kill soil-borne pathogens.
3) In the soils amended with 1% of biochar, the pharmaceutical residues in the radish roots showed three patterns in comparison with ________ of the unamended soil.
4) The volume of the sun is 1,3000,000 times ________ of the earth.

Task 2 Read the following sentences and identify the reference words.

1) Once it is combined, nitrogen is not chemically inert any longer.
2) Tobacco is grown from April to October, and in this process, the rapeseed is subsequently grown from October to March of the next year.
3) The Brassicaceae family produces sulfur compounds that break down to produce isothiocyanates that are toxic to many soil organisms.

Task 3 Identify the cohesive ties in the following Introduction and write them in the corresponding box.

Title: *Impact of 1-methylcyclopropene Treatment on the Sensory Quality of "Bartlett" Pear Fruit*

Introduction:

The demand for high-quality fruit in the marketplace has increased in recent years (Harker et al., 2003; Jaeger and Harker, 2005; Zhang et al., 2010). To fulfill consumer expectations and encourage repeat purchases, fruits in the market must have reliably good sensory characteristics. European pear (Pyrus communis) fruit are often characterized by their intense flavor and juicy-buttery texture which are important attributes of good eating quality (Rosen and Kader, 1989; Murayama et al., 1998). Previous studies have established that consumers are willing to pay a premium price for pears that are available ripe or close to ready-to-eat (Jaeger et al., 2003; Zhang et al., 2010). These studies also indicated that the most common reasons for liking pears were sweetness and juiciness, while firm gritty textures and lack of flavor were disliked. The produce industry is increasingly responding to consumer-driven demand and would like to provide high-quality pears over an extended marketing season (Winfree et al., 2004).

European pears naturally ripen in association with a climacteric rise in rates of ethylene production (Hansen, 1943). This ripening capacity is related to the physiological maturity of fruit at harvest (Chen and Mellenthin, 1981). For example, "Bartlett" pears picked early in the season produce low levels of ethylene at harvest, ripen very slowly, and consequently lack characteristic flavor and texture (Pug et al., 1996; Villalobos-Acuna and Mitcham, 2008). "Bartlett" fruit harvested later in the season (firmness of 76 N) synthesizes higher rates of ethylene at harvest, ripens more rapidly, and develops a more intense flavor. To stimulate rapid and uniform fruit ripening, early-season pears are typically exposed to low temperature (0–10 ℃) or ethylene (100 μl/L) conditioning treatments (Agar et al., 2000; Mitcham and Thompson, 1996; Mitcham et al., 2000; Villalobos-Acuña and Mitcham, 2008) prior to ripening. The influence of these conditioning treatments on sensory quality has been studied (Makkumrai et al., 2014); however, there is limited published information on the effects of conditioning treatments on the sensory quality of pears harvested at different maturity stages.

Cohesive ties					
Grammatical cohesion			Lexical cohesion		
Conjunction	Reference	Substitution	Lexical reiteration: repetition, synonym, superordinate	Lexical chains	General nouns

3F Assignments: *Reading Comprehension and Vocabulary*

Part one Reading comprehension

Task 1 Read the following text and tell whether the statements that follow are true (T) or false (F).

biochar *n.* 生物炭
postulation *n.* 假设
Terra preta 亚马逊黑土
feedstock *n.* 原料（指供送入机器或加工厂的原料）
manure *n.* 肥料，粪肥
pyrolysis *n.* [化学] 热解；高温分解

The renaissance of research on soil application of **biochar** was initiated by the **postulation** of its role in the sustained fertility of Amazonian soils known as "**Terra preta**" and the recognition of its stability in soil, which results in a net reduction of atmospheric CO_2 (Lehmann, 2007). A range of agricultural and organic materials can be used to generate biochars with different characteristics (Spokas et al., 2011). **Feedstock** characteristics and thermal conditions affect the biochars' physical and chemical characteristics (Antal and Gronli, 2003; Singh et al., 2007; Cao et al., 2011; Cantrell et al., 2012; Novak et al., 2014). Generally the higher the thermal conditions, the higher the inorganic nutrient content except for N (Novak et al., 2012). Furthermore, **manures** are nutrient-rich feedstock materials and the **pyrolysis** of manures produces more nutrient-rich biochars than plant-based biochars (Sheth and Bagchi, 2005; Chan et al., 2008; Gaskin et al., 2008; Ro et al., 2010; Cantrell et al., 2012). However, environmental impacts, such as potential water pollution from adding these manure-based biochars to soil, are not clear at this time.

______ 1) This section of Introduction has the definition of biochar.

______ 2) This Introduction starts with the purpose.
______ 3) This Introduction is organized in general-specific pattern.
______ 4) Biochar is a key word in this section.
______ 5) The research gap is involved in this section.

Task 2 **Write an Introduction for a research you have been involved in.**

Part two Vocabulary

Collocation refers to the pattern of words that typically go together, such as "make a decision" and "key factors". These collocations may involve different combinations of nouns, verbs, adjectives, adverbs, or grammatical words such as prepositions. Having a good command of collocations is particularly helpful in building language and developing fluency, flexibility, and naturalness.

Task 1 **Choose the ONE word that CANNOT collocate with the identified word or phrase.**

1) ________ knowledge
a. construct b. gain c. disseminate d. learn

2) chronic ________
a. indigestion b. grumbler c. experiment d. diagnosis

3) signal ________
a. change b. discontent c. assignment d. failure

4) complex ________
a. belief b. personality c. calculations d. issues

5) ________ visibility
a. increased b. limited c. low d. perceptive

6) commit ________
a. fraud b. treason c. syndrome d. violence

7) ________ obligation
a. fulfill b. transmit c. undertake d. impose

8) stimulate ________
a. consumption b. employment c. measure d. escalation

9) loose ________

a. alliance b. morals c. translation d. speech

10) assume ______

a. office b. role c. idea d. strategy

Task 2 Match the verbs listed in the box with the nouns 1)–5).

develop	do	have	make	put

1) ________ a link / a contribution / an impact / an analogy / an investment / assumption
2) ________ research
3) ________ the capacity / an impact
4) ________ intelligence / a theory / a concept
5) ________ an emphasis

Task 3 Complete this paragraph with the words listed in the box.

showed	apparent	significant	described	confirmed	indication

Given that it is not possible to differentiate cells in the desiccated (干燥的) layer, a better 1) ________ of the thickness of both the desiccated and lignified (木质化的) layers could be obtained by measurement using a graticule (方格图). For the lignified layer, highly 2) ________ differences among cultivars were observed for 2, 4, 6 and 10 days of healing. Results were less clear for the desiccated cell layers, but cultivar differences were significant after 4 days and highly significant after 10 days. Roots that had been stored for 1 and 6 weeks were used, but a significant effect of storage time ($p < 0.001$) was only 3) ________ after 10 days of healing, at which time SPK004 and Kemb10 had much thinner lignified layers, and thicker desiccated layers, while the other three cultivars 4) ______ no change. The cultivars SPK004 and Kemb10 had the thinnest lignified layers throughout healing; some roots of these cultivars completely failed to lignify. Yan Shu 1 and Zapallo had the thickest lignified layers. The results 5) ________the findings of Walter and Schadel (1983) that lignification is complete after four days, but disagreed with the results of St Amand and Randle (1991), who 6) ________an almost linear increase in the number of lignified cell layers from 0 to 7 layers between 3 and 12 days after wounding (at RH 85% T 29 °C).

Unit 4 Methods

Main Contents	Learning Objectives
Subgenre: *Methods*	★ Understanding the components of the Methods. ★ Learning to write the Methods section.
Organization: *General-specific*	★ Understanding the general-specific organization in the Methods.
Rhetorical Function: *Process*	★ Analyzing written description of processes. ★ Recognizing different types of processes. ★ Learning to write a paragraph describing a process.
Register: *Passive voice*	★ Understanding passive voice. ★ Identifying passive voice. ★ Learning to write an instruction/process using passive voice.
Cohesion (3): *General Nouns*	★ Understanding general nouns. ★ Recognizing general nouns in paragraphs and texts. ★ Learning to use general nouns to summarize ideas.
Assignments: *Reading Comprehension and Vocabulary*	★ Reading: Understanding the Methods better through in-depth reading. ★ Vocabulary: Learning to use adverbials for cohesion.

4A Subgenre: *Methods*

Although the Methods section typically follows the Introduction, researchers often write this section first. It is usually the easiest section to write and it answers some of the most basic questions that researchers need to deal with at the beginning of their research.

1. Elements of Methods

The Methods section serves as a "how-to" manual which describes how the research was carried out with technical details. This section may be brief with just one or two paragraphs, or quite long with subheadings when careful elaboration is necessary. Either way, it is usually composed of four basic elements (Cai, 2020): 1) the data or sample, 2) the materials, 3) the procedure, and 4) the data analysis.

Table 4.1 Elements of the Methods section

Subsections	Moves
Research design	Providing overview and/or research setting
Data or sample	Move 1: describing **data sources**, such as the data size and location, the sample population, its characteristics. Move 2: introducing **the criteria** for data collection or the sampling technique, and data collection procedure. Move 3: indicating **the advantages or restrictions** (e.g. representative) in choosing the sample.
Materials	Move 4: describing **research apparatus** such as what materials like equipment and tools you used to run your experiment or surveys and to acquire data, including the source of materials.

(To be continued)

(Continued)

Subsections	Moves
Procedure	Move 5: recounting **experimental process**, giving the reader a summary of each step in the execution of the research or experiments. Move 6: describing **experimental variables**. Move 7: **justifying the methods and procedures** employed.
Data analysis	Move 8: **evaluating and categorizing the data** collected from experiments or surveys by using some data analysis method. It might define terminologies, introduce analytical instrumental procedure and justify analysis procedure.

The Methods section should be concise, precise and logical with sufficient details for readers to evaluate the validity and reliability of your results and conclusion. If the method has been previously used by other researchers, remember to cite the original work.

Example 1

Title of the paper: *Aphicidal Activity of Bacillus Amyloliquefaciens Strains in the Peach-potato Aphid (Myzus persicae)*

Materials and Methods

Aphids

[Data or sample] Myzus persicae (Sulzer) was reared on radish (Raphanus sativus L). Aphids used in the experiments came from a colony maintained at the Faculty of Natural Science (National University of Salta, Salta, Argentina). This colony was initiated from a single virginoparous apterous individual collected in field in 2009. The colony was reared in a climate chamber at 22 ± 2 °C, 30%–40% R.H., and 16/8 h light/dark photoperiod to induce parthenogenesis. A new colony was started every week, and newly moulted apterae adult aphids were used for the experiments.

[Materials] Microorganisms, suspensions of bacterial cells (CS), heat-killed cell suspension (hkCS), cell-free supernatants (CFS), and isolated lipopeptide fractions (LF)

Aphicidal assays with B. amyloliquefaciens strains

[Procedure] The biological effect of B. amyloliquefaciens strains on M. persicae were assessed by offering the insects with artificial diets supplemented with CS, CFS, hkCS or FL, and after 4 days the adults mortality and the number of nymphs, alive and dead, were counted. The aphid's artificial diets contained 150 mM amino acids, 500 mM sucrose, vitamins, and minerals and were administrated through parafilm sachets on diet cages as previously described (Fig. 1) (Machado-Assefh et al., 2015; Prosser and Douglas, 1991). Diet cages consisted of plastic cylinders (3 cm diameter and 2 cm high) sealed on top with a diet sachet containing 100 μL of diet solution between two layers of parafilm, and sealed at the bottom with a mesh (Koga et al., 2007; Prosser and Douglas, 1991). Five recently moulted apterae adult aphids were placed on the diet cages and were maintained in a climate chamber at 22 ± 2 ℃, 30%–40% R.H., in darkness. After 4 days of treatment, the number of adults and offspring, dead or alive, were counted per cage.

Five assays were carried using 100 μL of artificial diet, each with an added portion of the different B. amyloliquefaciens strain derivatives, as follows: 1) 10 μL of CS from strains CBMDDrag3, PGPBacCA2, or CBMDLO3, on days 1, 4, and 6 after incubation; controls were plain diet and diet with 10 μL of peptone water; 2) 10 μL of CFS from strains CBMDDrag3, PGPBacCA2, or CBMDLO3, on days 1, 4, and 6 days after incubation; controls were plain diet and diet with 10 μL of LB medium; 3) 1, 5 and 10 μL CFS from strains CBMDDrag3 and CBMDLO3 on day 4 day after incubation; control was only plain diet; 4) 10 μL of CS and hkCS from strain CBMDLO3 on day 4 after incubation; controls were plain diet and diet with 10 μL of LB medium; 5) LF from CBMDDrag3 at a concentration of 5.8, 14.5, 29 and 58 μg of LF on 100 μL of diet, and LF from CBMDLO3 at 5, 12.5, 25 and 50 μg/100 μL of diet; controls were plain diet and diet plus 10 μL of distilled water at pH 8.

Assay 1) was repeated twice, and assays 2), 3), 4) and 5), were run in triplicate; in all cases, they were carried out under the same conditions and using the same solutions of CS, CFS, hkCS, and LF. The final number of replicates per assay 1) was n = 10, whereas for assays 2), 3), 4) and 5) the number of replicates varied between n = 10–15.

Statistics analysis

[Data analysis] The aphicidal effect of B. amyloliquefaciens strains on M. persicae, based on mortality of adults, number of nymphs, and total number of nymphs alive and dead, was analyzed by the Kruskal-Wallis non-parametric analysis of variance at one way of classification, followed by multiple comparisons between means of treatments with Bonferroni correction (Weisstein, 1999). Analysis were conducted with InfoStat 2015 (Di Rienzo et al., 2015).

Example 2

Title of the paper: *Influence of Genetically Modified Soya on the Birthweight and Survival of Rat Pups*

Methods

Animals

[Data or sample] Wistar rats were used as the subjects in the experiment. The animals were brought up to sexual maturity on laboratory rat feed. When their weight reached about 180–200 g, the female rats were divided into 3 groups, housed in groups (3 rats/cage), and kept under normal laboratory conditions. The feeding scheme was as follows. Females in every cage daily received dry pellets from a special container placed on the top of their cage. Those rats receiving soya flour supplement were given the soya flour in a small container placed inside their cage (20 g × 40 ml water) for three rats, so 5–7 g flour for each rat every day.

Experiment

[Procedure] One group of female rats of 180–200 g in weight was allocated to the experimental group, and received 5–7 soya flour/rat/day prepared from Roundup-Ready soya, added to the rat feed for two weeks. Another group females (3) were allocated to the control group, but their diet was supplemented with the same amount of soya flour, prepared from the traditional soya in which only traces (0.08 ± 0.04%) of the GM construct were present, most likely resulting from cross-contamination. We also introduced a positive control group (in two cages: 3×3), which had not been exposed to soya flour. Therefore females only got the standard laboratory feed without any supplementation, although it is acknowledged that the energy and protein content of this diet was less than in the other two groups.

After two weeks on the diets, all groups of 3 females were mated with two healthy males of the same age, which had never been exposed to soya flour supplements. In order to avoid infection of females, the sperm count and quality had not been determined. We carried on feeding the respective diets to all females during mating and pregnancy. Upon delivery, all females were transferred to individual cages, and the amount of soya supplement was increased by an additional gram for every pup born. Lab feed and water were available for all animals during the experimental period. When the rat pups opened their eyes and could feed themselves (from 13–14 days of age), the daily dose of soya supplement was increased to 2–3 g for every pup, although all rats had free approach to the soya. All rats ate their soya portions well. After the experiment was finished the organs of some pups were taken out and weighed.

[Data analysis] The level of mortality was analyzed by the one-way ANOVA, using the Newman-Keuls test for share distribution. The pup's weight and its distribution were checked by Mann-Whitney test and Chi-square in StatSoft Statistica v6.0 Multilingua (Russia).

Task 1 **Read the following edited Methods section from the paper entitled "*Fish Invasions in the World's River Systems: When Natural Processes Are Blurred by Human Activities*" closely and identify four elements in the Methods section. Mark the number in the text after the corresponding element.**

Materials and Methods

① We conducted an extensive literature survey of native and non-native freshwater species check lists. ② Only complete species lists at the river basin scale were considered, and we discarded incomplete check lists such as local inventories of a stream reach or based only on a given family. The resulting database was gathered from more than 400 bibliographic sources including published papers, books, and grey literature databases (references available upon request). Our species database contains species occurrence data for the world's freshwater fish fauna at the river basin scale (i.e. 80% of all freshwater species described and 1,055 river basins covering more than 80% of the Earth's surface). It constitutes the most comprehensive global database for freshwater fish occurrences at the river basin scale and, to our knowledge, the largest database for a group of invaders. We considered as non-native a species (i) that did not historically occur in a given basin and (ii) that was successfully established, i.e. self-reproducing populations. Estuarine species with no freshwater life stage were not considered in our analyses.

③ We first mapped the worldwide distribution of (i) the non-native species richness per basin and (ii) the percentage of non-native species per basin (i.e. the ratio of non-native species richness / total species richness). We used three classes of percentage (Figure 1A) and richness (Figure 1B) of non-native species to draw colour maps. Other maps with more classes were tried and provided similar results. We selected the one that minimised differences in sample size (i.e. number of river basins) between classes. The percentage of non-native species per basin was used to define invasion hotspots where more than a quarter of the species are non-native (i.e. the third class of percentage of non-native species; red areas in Figure 1A). For each of the three levels of fish invasion ([0%–5%], [5%–25%], [25%–95%]), we determined the percentage of species facing a high to extremely high risk of extinction in the wild, i.e. the vulnerable, endangered, and critically endangered fish species according to the IUCN Red List. The list of basins for the three levels of invasion is provided in Dataset S1.

④ In our dataset, the seven environmental and human variables are not independent (Pearson's correlation coefficient values ranging from −0.25 to 0.79, Table S1). We therefore evaluated the independent explanatory power of each environmental and human variable by using hierarchical partitioning [26–28, 51, 52], a method based on the theorem of hierarchies in which all possible models in a multiple regression setting are considered jointly to attempt to identify the most likely causal factors (explanatory approach).

⑤ Hierarchical partitioning does not provide information on the form of the relationship (positive or negative) between the number of non-native species and each explanatory variable. To test the "human activity" hypothesis, we analyzed the form and the significance of the relationship between each variable related to the "human activity" hypothesis (GDP, percentage of urban area, and population density) and the residuals from a GLM with a Poisson error. Then, to test the "biotic acceptance" hypothesis, we analyzed the form and the significance of the relationship between each variable related to the "biotic acceptance" hypothesis (i.e. number of native species, altitudinal range, basin area, and net primary productivity) and the residuals from a GLM explaining the number of non-native species by using the human activity-related variables (i.e. GDP, percentage of urban area, and population density). Lastly, to test the "biotic resistance" hypothesis, we analyzed the form and the significance of the relationship between the number of native species and the residuals from a GLM explaining the number of non-native species by using independent variables related to the "biotic acceptance" and "human activity" hypotheses (i.e. altitudinal range, basin area, net primary productivity, GDP, percentage of urban area, and population density). This allowed us to control for the effects of environmental conditions, propagule pressure and habitat disturbance. ⑥ To test the relationship between the model residuals and each explanatory variable, we performed a Spearman rank correlation test, because the model residuals were not normally distributed.

a. Data or sample: ________ b. Materials: ________

c. Procedure: ________ d. Data analysis: ________

Task 2 Maryam Najafabadi, an Iranian teacher at Islamic Azad University planning a research into agricultural marketing, talks about how he will conduct his research. The provisional title for his RP is *Agricultural Marketing Challenges and Barriers in Iran*. Read his account, and write the Methods section.

Maryam Najafabadi: The purpose of this research is to investigate how agricultural experts perceive the barriers to agricultural marketing. I have found 134 agricultural experts who are willing to take part in my survey mainly in Tehran province. I'm going to collect my data by distributing a questionnaire to the agricultural experts. I'll develop the questionnaire items on the basis of previous literature. My friend, who is an expert on designing questionnaire, will help examine the validity of the research model. I decide to use a five-point Likert scale for the measurement with 1 being strongly disagree and 5 being strongly agree. To prove the reliability of the instrument, I'll choose 15 experts randomly from the target population to take a pretest. (Cronbach's alpha turned out to be 80%.)

2. Sentence structures frequently used in Methods

The following sentence structures are frequently used in describing methods. They may help you identify each element in this part or write an effective Methods section.

Table 4.2 Sentence Structures Frequently Used in Methods

Elements	Sentence structures
Data or sample	... was designed to collect data for the study. The experiment was conducted under the condition of ... ... method had been employed to obtain ... in the samples.
Materials	The questions used in our studies are adapted from the studies by Richards. ..., based on the classical approach, was designed for ... It was designed/classified ... The use of ... to simulate ... was highlighted.

(To be continued)

(Continued)

Elements	Sentence structures
Procedure	We examined/counted/included ... We first mapped ... Then we selected ... Lastly, to test the hypothesis, we analysed ...
Data analysis	We analyzed the data with a number of tests. The level of ... was analyzed by ..., using the test ...

Task 3 **Read Example 1, Example 2 and the text in Task 1 above again and find out the language which can help you to identify the elements.**

	Data or sample	Materials	Procedure	Data analysis
Example 1				
Example 2				
Text in Task 1				

4B Organization: *General-specific*

It's possible that the development of the Methods part begins with a general sentence, followed by more details explaining how the study/research was conducted in the latter part of the paragraph, i.e. a typical GS pattern.

Example 1

Title of the paper: *Effects of Coastal Fish Farms on Body Size and Isotope Composition of Wild Penaeid Prawn*

① A small pilot feeding trial was carried out in captivity in order to obtain baseline isotopic values of prawns fed exclusively on fish feed. ② A total of 20 prawns were taken alive from the fishing boats to the laboratory at the University of Alicante on three occasions, where they were distributed equally between two 100 L. glass aquaria filled with artificial seawater (37 ppm) and kept at 20 °C with continuous aeration. ③ Prawns were fed four food pellets/day/individual, being the same type (Skretting l-Alterna) as that used in the commercial fish farms. ④ Uneaten food pellets were siphoned daily before each feeding and approximately half of the water was replaced once a week to maintain water quality. ⑤ Any dead prawn were removed within the next 12 h, but these were not replaced during the experiment. Individuals were measured, weighed and stored at −80 °C until analysis, and those individuals which could be individually identified and survived for 4–55 days were used for the analysis of the effects of feeding duration on isotopic values (n = 10). ⑥ Feeding was ascertained by visual observation and presence of digested fish food in the digestive tract following dissection. ⑦ To estimate the likely isotope values of prawn feeding on fish pellets in the wild, we employed a trophic discrimination factor (TDF) of 2.25% for δ^{15}N and 0.25‰ for δ^{13}C based on equations for invertebrate tissue (Caut et al., 2009), and compared these to empirical values derived from the feeding trial in captivity.

Sentence ① presents a general statement of the method adopted in the experiment.

Sentences ②–⑥ describe the specific procedures.

Sentence ⑦ states the tools and materials used to obtain the data.

A general statement of the method or approach used in the experiment helps readers better evaluate the subsequent procedures conducted.

Example 2

Title of the paper: *Wheat Improves Nitrogen Use Efficiency of Maize and Soybean-based Cropping System*

① Statistical analyses were performed using SAS statistic software (Statistical Analysis System, version 9.3, SAS Institute, NC, USA). ② Residuals were found homogeneous, normal-distributed using the Shapiro-Wilk W test (P = 0.98), and no significant outliers were detected by Lund's test. ③ Mixed models were used for analysis of variance, with crop rotation, tillage system and N as fixed effects, and year and replication as random effects. ④ PROC NLIN with Marquardt iterative method was used to fit crop yield response to N rate to quadratic plateau models. ⑤ Models were constrained such that the linear regression coefficient was greater than or equal to 0, and the quadratic coefficient was less or equal to 0 (Gaudin et al., 2014). ⑥ Comparison of predicted values from regression curves was based on a t-test using standard errors obtained from the Mixed model. ⑦ Type I error rate was set at 0.05 for all tests.

Sentence ① introduces the statistic software used to analyse the data.

Sentences ②–⑦ provide detailed information of using the statistic software to analyze the data.

Task 3 **The following sentences are from the Methods of an essay entitled *Impact of Different Crop Geometry in Maize on Fall Armyworm*. Put the sentences in the correct order, from general to specific information.**

① Each treatment was replicated four and each replication data was the mean of data obtained from 20 plants.

② The observations on larval population (number of larvae/plant), leaf damage (damage severity) and percentage of plant infestation along with yield were then recorded.

③ The present investigation was to determine the effect of different plant spacings against fall armyworm and the yield of maize.

④ Field trial was carried out during the year rabi, 2019 and 2020.

⑤ Three crop spacings were selected based on (i) precision farming (75 cm × 20 cm), (ii) TNAU recommended spacing (60 cm × 25 cm) and (iii) farmer's practice (45 cm× 30 cm) was maintained by hand dibbler having a net plot size was 7 m × 2 m were evaluated against fall armyworm incidence and damage in comparison with check.

⑥ Three different crop spacings were evaluated under three two hybrid maize varieties.

⑦ Finally, the yield data were recorded for each treatment separately and grain yield was expressed in kg/ha.

Correct order: ________________

4C Rhetorical Function: *Process*

Of the three rhetorical functions discussed in Unit 1, i.e. to describe, to explain and to persuade, observing and then describing a process is a less challenging task. However, texts are almost always multifunctional because writers may have different purposes to achieve through different rhetorical functions.

A process is a series of steps or actions taken to achieve a particular end or objective. It can be a series of activities, methods, procedures, or systems that work together to accomplish a task. The description of the process usually includes materials or equipment used, as well as any specific conditions or parameters that were observed during the process. It also includes the data or observations that were collected during the process, and any analysis that was performed on the data.

Overall, the description of a process in a research paper should be clear, concise, and well-organized, with all relevant information and data presented in a logical and easy-to-follow manner.

Example 1 is an excerpt from the Methods section of a research article which describes how an experiment was conducted with materials, procedures, equipment and analysis performed on the data.

Example 1

Title of the paper: *Impact of 1-methylcyclopropene Treatment on the Sensory Quality of "Bartlett" Pear Fruit*

[Materials] As pears of each treatment approached eating ripe, fruit were randomly selected and assessed daily for flesh firmness as described above. [Procedures] Five representative pears from each treatment that exhibited firmness of 13±2 N were prepared for instrumental and sensory analyses. Briefly, each fruit was cut into twelve 5 mm thick longitudinal wedges from the stem to blossom end. The core was carefully removed. A total of forty-five slices (five pears × nine remaining slices) per treatment were evaluated by the panelists for sensory quality. Two fruit slices per treatment were prepared for each panelist. Pear slices were served with a cup of water and unsalted crackers as palate cleansers. [Equipment] The sensory evaluation software Compusense Five (Compusense, Ontario, Canada) was used in all training and tasting sessions.

[Analysis performed on the data] The accuracy and validity of the sensory panel was checked after each tasting session by ANOVA (P = 0.05), with the statistical differences between the means determined by LSD. All the descriptors were validated for their sample effect, including overall aroma intensity, apple, pear, green, and fermented aroma; firmness, juiciness, and grittiness; and sweet, tart and fruity taste.

Some generic words can be especially useful in describing a process as shown in Example 2 in bold. They provide readers a framework to understand a process or procedure or to reduplicate it.

Example 2

The traditional process for making soy sauce **is still used** for making sauce of high quality. In Japan **it starts** in April **and continues** for a whole year, making use of the changing temperatures in the different seasons. **There are several stages,** and fermentation is carried out by many different moulds, bacteria, and yeasts which **successively** predominate in the developing sauce as conditions change to suit them. **In outline, the process is this**. Defatted, steamed soya beans and roasted, crushed wheat are mashed together. The mixture is inoculated with tane koji, a starter culture of the two necessary *Aspergillus* moulds, and is allowed to ferment, **then** mixed with a strong salt solution and inoculated with another starter containing several kinds of bacteria and yeasts **for a further fermentation** which lasts from 8 to 12 months. **The reactions in this last period** create a complex blend of substances contributing to the final flavour. The chief elements are salt, amino acids, organic acids (lactic and acetic), alcohols, sugars, and numerous volatile aromatic substances including vanillin, the flavour principle of vanilla. **When** fermentation is complete, the mixture is filtered or racked to extract the sauce; **and this is commonly** pasteurized to kill the remaining organisms and arrest fermentation. One ton each of defatted soya beans, wheat and salt **produces** 5,000 litres of soy sauce.

Task 1 **Identify the words or phrases in the following text that you can use to describe any process.**

Photosynthesis is a process by which green plants and some other organisms convert light energy from the sun into chemical energy in the form of glucose. The process of photosynthesis can be divided into two stages: the light-dependent reactions and the light-independent reactions.

The first stage is light-dependent reactions which take place in the thylakoid membrane of the chloroplasts. The process begins when light energy is absorbed by chlorophyll, the green pigment in plants. This energy is then used to convert water molecules into oxygen gas (O_2), protons (H^+), and electrons (e^-). The oxygen gas is released into the atmosphere, while the protons and electrons are used in the next stage of photosynthesis.

The second stage, light-independent reactions, also known as the Calvin cycle, take place in the stroma of the chloroplasts. The process begins when the protons and electrons from the light-dependent reactions are used to convert carbon dioxide (CO_2) into glucose ($C_6H_{12}O_6$). This process requires a series of chemical reactions that involve several enzymes. The glucose produced in this stage can be used by the plant for energy or stored for later use.

Overall, photosynthesis is a complex process that involves the absorption of light energy, the conversion of water and carbon dioxide into glucose, and the release of oxygen gas. The process is essential for the survival of plants and other organisms that depend on them for food and oxygen.

Task 2 **Work with a partner and write a short description of approximately 100 words on the planting of corn, based on the notes below. You will need to identify the nouns in the notes and convert them to verbs in your description. Then, compare your description with other students.**

a. Selection — choosing a type of corn

b. Preparation — preparing your garden, including a location with good sunshine and soil that is nitrogen rich and well manured

c. Planting — planting 2–3 seeds together at each placement to increase the odds of the seed germinating
d. Watering — applying about one inch of water to the base of the plants each week
e. Weeding — keeping the corn weed-free until it's about knee high
f. Waiting — waiting until the corn finishes growing
g. Harvesting — picking your corn when you see a dry, brown silk tail at the top of the ear

4D Register: *Passive Voice*

The passive voice is often used in academic texts because it helps the author: 1) focus on the topic of the essay or article; 2) be objective; and 3) avoid saying who did the action because it is obvious, unnecessary, or unimportant.

Examples

1. GM plants are highly regulated ...
2. Water tables are increasingly being decreased ...
3. Farm workers have been poisoned by pesticide, ...
4. The mixture is filtered or raked to extract the sauce ...

The subjects of the sentences in the "Examples" are *GM plants*, *water tables*, *farm workers* and *the mixture*. The author uses the passive voice to focus on these topics. It is unnecessary or unimportant to say who does these things as they are not the focus of the description. Since the passive voice emphasizes the idea, the action, the object or the event rather than the person who carries it out, it seems more objective.

In Methods, where materials, tools, data, objects, procedures, etc. are the focus of the description, the passive voice is more frequently used. It highlights the research itself and helps the reader to focus on what they are interested in.

Task 1 **Read the following sentences, identify the topic (the main focus of the sentence), find the main verb and tell whether it is in the active or passive voice.**

1) You can see signs of change in this environment.
2) The natural balance can be damaged by farming which is the main human cause to the process of desertification.

3) A third of the world's land surface is currently affected by desertification.
4) Human actions are worsening natural global warming and climate change.
5) Desertification can be caused by using too little or too much water.

Task 2 Change the voice of the sentences in Task 1. A model and the opening phrases have been given.

Model: We conducted this study at Massey University's Tuapaka farm 15 km east of Palmerston North, New Zealand. (active)
This study was conducted at Massey University's Tuapaka farm 15 km east of Palmerston North, New Zealand. (passive)

1) Signs of change ______________________________.
2) Farming ______________________________
______________________________.
3) Desertification ______________________________.
4) Natural global warming and climate change ______________________________
______________________________.
5) Using too little or too much water ______________________________.

Task 3 Correct the mistake in each of the following sentences. Then decide whether these sentences are in the active or passive voice.

1) All manipulations approved by the Massey University Animal Ethics Committee.
2) Many companies have accused of trying to obtain a monopoly to limit competition.
3) It expects that nanotechnology will provide many benefits for society.
4) Millions of dollars have spent on researching thousands of medical procedures and drugs.
5) To date no studies have been examined the use of herb-clover mix for early weaned lambs.
6) Lambs and ewes allocating to treatments at L58 in 2015 and L51 in 2016 were rotationally grazed.

Task 4 Complete the following paragraphs with the words in brackets in the proper form.

Over-grazing. Marginal grassland 1) ________ (have) a sustainable carrying capacity —

the number of animals that 2)________ (can support) without causing long-term damage. If this number 3) ________ (exceed), the system becomes unsustainable and the vegetation and soil 4) ________ (deteriorate). If this 5) ________ (continue), it can lead to desertification.

Over-irrigation. If plants 6) ________ (appropriately irrigate), little water should be wasted. However, if land 7) ________ (over-irrigate), salinization can 8) ________ (occur). This creates an impermeable and infertile salty crust on the surface, which (according to UNESCO) is a key feature of desertification. Other human activities that 9) ________ (can damage) the soil and vegetation (leading to soil erosion and ultimately desertification) include: road building, deforestation, and inappropriate tourism.

4E Cohesion (3): *General Nouns*

On the borderline between grammatical and lexical cohesion is the class of general nouns, a small set of nouns having generalized reference within the major noun classes, such as those referring to person, fact, place and the like.

General nouns are important for the following reasons:

- They are very common in academic texts and in the collocations found there.
- They are one of the main ways that academic writers link and organize ideas in their texts.
- They are applicable to a wide range of subjects and are thus useful for students to learn.
- They help student writers to summarize texts.
- When students can use general nouns, their writing begins to appear more competent and academic to subject specialists.

In all, general nouns are really useful words because they summarize and classify the person, thing, idea or event which has been described in more detail in previous text. Writers can use general nouns to show their stance by referring to the information they present, e.g. a problem, an issue, a solution, an advantage or a success.

Example

Title of the paper: *Rural Development Potential in the Bioeconomy in Developed Countries: The Case of Biogas Production in Denmark.*

In recent decades, depopulation, loss of workplaces, deterioration of the housing stock, and loss of private and public services, etc. are the interconnected **challenges** that rural areas in Western countries have faced. To counter these **problems**, the bioeconomy — a biomass-based economy — has often been mentioned as a promising development opportunity.

Policymakers hope that bioeconomy can promote the economic growth and job creation in rural areas in developed countries and they have expressed much optimism about the **potentials**.

An increase in the biogas production by 10% in Denmark has given a permanent increase of 342 jobs and an extra annual income of approximately 21 million euros. This **success** will enhance the development and economic growth of the rural areas.

Consequently, if all available biomass from farm manure were to be used in biogas production, it would result in 3,420 jobs, and the **expansion** would reduce fossil fuel dependence and improve economic and environmental sustainability of primary production.

Experts also believe that the **improvement** of bioeconomy in Denmark can further boost local rural economies through increased investment in skills, knowledge, innovation and new business models.

In the text above, the general nouns are highlighted in bold, and the content they refer to are underlined. As the example shows, the general nouns in cohesive function are almost always accompanied by *the* or other determiners, e.g. possessive adjectives like *its* and *their*, and demonstrative adjectives like *this/these* and *that/those,* which tell the reader that they refer to other parts of the text.

Task 1 Underline the general nouns and the part it refers to in the following text.

Various attempts have been made to explain where intelligence comes from. In the nineteenth and twentieth centuries, the essential argument of heredity versus environment emerged. These terms are often known as "nature" and "nurture" respectively. Essentially the arguments are concerned with the extent to which intelligence is inherited through the genes a person is born with (heredity or nature) or formed through a person's life and their surroundings (environment or nurture).

These theories led to the concept of how to measure intelligence. Early attempts at measuring intelligence (e.g. Galton, 1869) associated it with social inequalities. According to this point of view, the role of environmental factors had to be recognized alongside the part played by heredity. This work led to the construction of IQ (intelligence quotient) tests to measure intelligence. The measurement of IQ originated in the work of the statistician Spearman (1904), who introduced the concept of "g" (general intelligence) to describe the general cognitive ability that he thought lay behind specific abilities and forms of intelligence (linguistic, mathematical, spatial, musical, etc.). The development of these tests seemed to promise the possibility that the relationship between material inequalities and social inequalities could be studied with mathematical precision.

General nouns	The ideas they refer to

Task 2 **Read the text in Task 1 again and find out the cohesive ties in the text.**

Various attempts have been made to explain where intelligence comes from. In the nineteenth and twentieth centuries, the essential argument of heredity versus environment emerged. These terms are often known as "nature" and "nurture" respectively. Essentially the arguments are concerned with the extent to which intelligence is inherited through the genes a person is born with (heredity or nature) or formed through a person's life and their surroundings (environment or nurture).

These theories, led to the concept of how to measure intelligence. Early attempts at measuring intelligence (e.g. Galton, 1869) associated it with social inequalities. According to this point of view, the role of environmental factors had to be recognized alongside the part played by heredity. This work led to the construction of IQ (intelligence quotient) tests to measure intelligence. The measurement of IQ originated in the work of the statistician Spearman (1904), who introduced the concept of "g" (general intelligence) to describe the general cognitive ability that he thought lay behind specific abilities and forms of intelligence (linguistic, mathematical, spatial, musical, etc.). The development of these tests seemed to promise the possibility that the relationship between material inequalities and social inequalities could be studied with mathematical precision.

(To be continued)

(Continued)

Cohesive ties					
Grammatical cohesion			Lexical cohesion		
Conjunction	Substitution	Reference	Lexical reiteration: repetition, synonym, superordinate	Lexical chains	General nouns

Task 3 Complete each sentence with a proper general noun from the list, and underline the part it refers to.

problems	growth	knowledge	situation	reasons	idea

1) Schema theory suggests that what we already know will influence the outcome of information processing. This ________ is based on the belief that humans are active processors of information.
2) Cognitive psychologists are involved in finding out how the human mind comes to know things about the world and how it uses this ________.
3) Psychologically, observation is less threatening, and in terms of cost, it is also likely to be cheaper than other methods. For these ________, observation is a very practical and effective assessment method for people at work.
4) Cybercrime and internet-related crime refer to crimes which take place online rather than in real life. These serious ________ include fraudulent financial transactions, crimes of a racial nature, and certain sexual crimes.
5) Athletes who set achievable goals typically improve their performance. The ________ is similar in education.
6) In the 1930s in the UK, it cost nearly a week's wages to make a three-minute phone call; today, it is around one penny to make a one-minute call from the UK to India. More recently, internet usage has exploded, from 100 million users worldwide in the early 1990s to more than a billion in 2006 (Webster and Hamilton, 2009). Were it not for the ________ and falling costs of telecommunications, then globalization is unlikely to have happened at the rate it has.

4F Assignments: *Reading Comprehension and Vocabulary*

Part one Reading comprehension

Task 1 Read the following text and tell whether the statements that follow are true (T) or false (F).

Material and Methods for the Preliminary Study

Litter sample were collected from a standard broiler farm in Georgia according to EPA method (EPA-821-B-04-006, 2004). All the **poultry** litter samples were **refrigerated** at 2 to 4 °C prior to use to avoid **microbial** decay. Samples were screened using Screen # 20 (0.85 mm) (USA Standard Testing Sieve, Fisher Scientific Company, USA) on an electrical sieve shaker (Model CL 5028, Soil Testing Inc, Evanston, ILL, USA) and sample sizes were reduced using ASTM D 6913 standard methods. Prior to the actual screening, the standard shaking period was determined for the poultry litter according to the same standard. After the screening, all the samples were dried at 45 °C for 24 hours in an oven (Isotempoven, Fisher Scientific). All the samples were **ground** to 4 mm (FRITSCH, Pulverisette, Industries Trane, Idar, Oberstein) followed by further reduction in **particle** size to less than 1 mm (Thomas Scientific).

Heating value was measured using ASTM D5865 standards using a Bomb **Calorimeter** (Model 1351, Parr Instrument Company (Parr), Moline, IL). Proximate analysis (ash, volatiles, fixed carbon and moisture content) was done

poultry *n.* 家禽
refrigerate *v.* 冷藏
microbial *n.* 微生物的
grind *v.* 碾压
particle *n.* 微粒，颗粒
calorimeter *n.* 热量计；量热仪

Thermo Gravimetric Analyzer 热重分析仪

according to ASTM D3174 using a **Thermo Gravimetric Analyzer** (Model TGA701, LECO Corporation, St. Joseph, MI). Samples of the raw poultry litter, its coarse and fine fractions, and raw pine chips were sent to the Soil, Plant, and Water Laboratory, 2400 College Station Road, Athens, GA for ultimate, **compositional** and mineral analysis.

compositional *adj.* 成分的

The Thermo Gravimetric Analysis (TGA) and Differential Scanning Calorimeter (DSC) analysis were performed by heating the three fractions of poultry litter and a sample of raw pine chips (in 70 μL **alumina crucible** for TGA, and 100 μL aluminum crucible for DSC) from 25 ℃ to 600 ℃ in three segments: 1) **isothermal** heating for a minute at 25 ℃, 2) heating at the rate of 10 ℃/minute from 25 ℃ to 600 ℃, and 3) Isothermal heating for a minute at 600 ℃ (Model TGA/SDTA 851 and DSC873, Mettler Toledo, Star Systems, Columbus, OH). An inert atmosphere and the removal of gases and **condensate** products were achieved using 50 cm^3/min of nitrogen. The output from TGA and DSC analyses were analyzed and differences in thermal decomposition behavior were explained by the data obtained from compositional, mineral, and ultimate analysis in the light of previous research works.

alumina *n.* 氧化铝，矾土
crucible *n.* 坩埚
isothermal *adj.* 等温的
n. 等温线
condensate *adj.* 凝缩了的

______ 1) This Methods section followed a general-specific pattern.

______ 2) The data of this research were collected from a standard broiler farm in Georgia.

______ 3) This Methods section clarified the experimental process.

______ 4) This Methods section evaluated and categorized the data by using some data analysis methods.

______ 5) Three research apparatuses are used in this research.

Task 2 Read the following text and answer the questions that follow.

Animals and diet

wether *n.* 阉公羊

Six spring born (February to March) Suffolk **wether** lambs, initial mean BW of 26.4 ± 0.7 kg, were used in this study.

They were purchased from a commercial sheep farm at 4 to 5 months of age on July 8, 2013. All lambs were **deworm**ed prior to use in the study. The animals were randomly divided into two groups (n = 3, each group) to be of similar mean BW (26.1 vs. 26.7 kg) and were allocated to one of two dietary treatments for 9 months feeding period. Except for **metabolism** trials by total collection, the animals were housed in **well-ventilated pens** under continuous lighting and had free access to water and mineral blocks during the feeding experiment.

deworm *v.* 去除（动物身上的）虫

metabolism *n.* 新陈代谢

well-ventilated *adj.* 通风良好的

pen *n.* 圈；围栏；畜栏

Experimental procedure and measurements

Feed refusals were weighed and recorded daily for each animal through the feeding experiment. During the metabolism trials, animals were transferred from pens and were kept in metabolism **crates** in a temperature controlled room (20°C) under continuous lighting, and total collections of feces and **urine** were conducted. Total urine was collected for estimating **urinary** N excretion and MBN supply to the small **intestine**. Urine was collected into the container containing 50 mL of 10% (v/v) H_2SO_4 to keep urine pH below 3. On the final day of total collection, wether lambs were blood-sampled by **jugular vein** catheterization (18G × 2.5, 0.95 mm internal **diameter**, TERUMO Corporation, Tokyo, Japan) using **syringes** containing sodium citrate solution (3.9% w/v).

crate *n.* 大木箱；板条箱

urine *n.* 尿

urinary *adj.* 尿的

intestine *n.* 肠

jugular vein 颈静脉

diameter *n.* 直径

syringe *n.*（皮下）注射器

Chemical analysis

The **fecal** samples were dried in a forced air oven at 60°C for more than 48 h. Samples of air-dried **feces**, feeds and orts were ground to pass through a 1 mm screen. Samples of feeds, orts and feces were determined for OM, crude protein (CP), and ether extract (EE) by AOAC (1984) standards. **Blood plasma** concentrations of **glucose** and blood urea nitrogen (BUN) were determined using commercial kits (Glucose C-test Wako and BUN B-test Wako; Wako Pure Chemical Industries, Osaka, Japan).

fecal *adj.* 粪便的

feces *n.* 粪便

blood plasma 血浆

glucose *n.* 葡萄糖

Statistical analysis

Data were analyzed by a two-way repeated measures

analysis of variance (ANOVA) to evaluate the effects of diet and feeding period on feed intake, DG, digestibility, N retention, ruminal MBN yield and plasma metabolites concentration. Data for daily feed intake and DG were summarized as mean values for each month of the feeding period. All statistical procedures were performed using SPSS 14.0 (SPSS 2006).

Questions

1) How many elements can you find in this Methods section? Please write down these elements.
2) What is the data source of this research?
3) How many research apparatuses are used in this Methods section? Please name three of them.
4) Underline the sentences that use passive voice in this Methods section. What is the advantages of using the passive voice?

Part two Vocabulary

Adverbials help to introduce supporting information and make a text cohesive by establishing a logical and coherent flow between ideas and sentences. Words like *for example*, *for instance* are used to introduce an example; for explanation, you can use *in other words* (or *i.e.*); *similarly* and *likewise* are useful when you try to make a comparison; to introduce an evaluation and signal your stance, you can use adverbials like *surprisingly*, *significantly*; finally, you may use *in brief*, *in conclusion* or *in short* to summarize or conclude at the end of a text.

Task 1 Complete the following sentences with the phrases or words given below as cohesion.

For instance	in conclusion	in other words
likewise	surprisingly	essentially

1) ________, lambs are conventionally weaned between 10 and 14 weeks of age onto grass-clover pastures in lamb production system in New Zealand.

2) Early weaning of lambs was advantageous for the ewe in 2015.
________, this was not found to be the case in 2016.
3) Nonetheless, different feed has no significant impact on ewe health;
________, no differences in ewe body condition score were observed in both years.
4) These traits of herb-clover mix have resulted in improved ewe performance, ________, liveweight, body condition and milk production; ________, there has been considerable growth in liveweight of pre- and post-weaning lambs.
5) ________, the effect of early weaning onto herb-clover mix on lamb liveweight gain was more apparent when pasture conditions were restricted.

Task 2 **Choose a proper word or phrase from the following box for each blank to make the following paragraph more cohesive.**

for example	additionally	significantly
in brief	similarly	in other words

Agriculture plays a vital role in our society. It not only provides us with essential food and nourishment but also contributes 1) ______ to the economy. Farmers across the globe work tirelessly to cultivate crops and raise livestock to meet the growing demand for food. 2) ________, advancements in technology have revolutionized the agricultural sector. For instance, the use of precision farming techniques, such as GPS and remote sensing, has significantly improved efficiency in managing crops and reducing resource wastage. 3) ________, the implementation of sustainable farming practices has gained momentum. 4) ________, farmers are increasingly adopting methods that minimize the use of harmful chemicals, protect soil health, and promote biodiversity. 5) ________, crop rotation and organic farming have shown promising results in preserving the environment while ensuring high-quality produce. 6)________, agriculture is experiencing great challenges, such as climate change and limited access to resources, but with continued research, innovation, and collaboration, we can strive towards a more sustainable and resilient agricultural system.

Unit 5 Results

Main Contents	Learning Objectives
Subgenre: *Results*	★ Understanding the structure of a Results, section. ★ Identifying the structure of a Results section.
Organization: *General-specific*	★ Grasping the general-specific organization in the Results.
Rhetorical Function: *Graph Description*	★ Understanding graph description. ★ Learning to describe the graph with proper words.
Register: *Long Sentences*	★ Understanding long sentences. ★ Identifying the basic sentence structure in long sentences. ★ Learning to write long sentences.
Cohesion (4): *Substitutions*	★ Understanding substitutions. ★ Recognizing different cohesive ties in an academic paper.
Vocabulary: *Reading Comprehension and Vocabulary*	★ Reading: Understanding the Results section better through in-depth reading. ★ Vocabulary: Learning to use words describing trends.

5A Subgenre: *Results*

The Results section is the core of an academic article. It reports the data from the experiments, highlights key findings from the data and makes comparisons. The Results section should be written in a concise and logical way.

1. The structure of a Results section

The Results section of a paper contains two parts: statement of the findings and data presentation. The statement part uses language to directly answer the research questions raised in the Introduction part. Data presentation is composed of tables, figures or illustrations, which report statistical information aiming to support the statement. The two parts are combined to answer the research questions. The following (Cai, 2020) are the elements of a results section.

Table 5.1 The Structure of Results

Element 1	Research questions/hypotheses (optional)	Stating the major research questions
Element 2	Location statement (required)	Locating the tables and figures where the results can be found
Element 3	Major results (required)	Providing major data to support the results or findings
Element 4	Specific data (required)	Providing specific data to support the results or findings
Element 5	Comments (optional)	Providing comments (evaluation, comparison and explanation etc.) on some of the results

Example 1

Title of the paper: *Evaluation of Physicochemical Properties and Sensory Attributes of Biscuits Produced from Composite Flours of Wheat (Triticum aestivum l.) and Potato (Solanum tuberosum l.)*

[Location statement] The result of the physical analysis of the biscuits produced from wheat and potato flour blends is shown in Table 1, which shows that the supplementation of various levels of composite flours has a significant effect on weight, diameter and thickness.

[Specific data] The highest weight was recorded in control (100% wheat) and the lowest was recorded with the supplementation of 50% potato with values of 8.26 g and 7.61 g respectively.

[Major results] There were no significant difference between the treatments, but 50% supplementation of potato flour showed significant differences at ($p < 0.05$).

[Comment] This is in agreement with the work of Muhammad et al., 2014 who reported on their evaluation of quality biscuits prepared from wheat and cassava flours, that is as supplementation of cassava flour increases, the weight decreases.

Example 2

Title of the paper: *Surface Chemistry and Germination Improvement of Quinoa Seeds Subjected to Plasma Activation*

[Major results] A general result of the plasma activation experiments was a germination improvement for the plasma treated seeds with some differences depending on the type of plasma and treatment time.

[Location statement] A compilation of percentages of germinated Quinoa seeds as a function of seeding time for increasing periods of plasma treatment times is gathered in Fig. 1.

[Specific data] Figure la shows that, in comparison with untreated seeds, those treated with RF plasmas for 10s experienced an improvement in germination, reaching almost 100% after five days. Plasma treatment times of 60s produced a germination success rate of 80% after 8 or more days. However, a considerable decrease in germination percentage was obtained for seeds treated for 180s, indicating that, as found by others, an excess of RF plasma might be deleterious for seed germination (see in supplementary information, Figure S1, the surface damage induced on seeds that were RF treated for long periods of time).

[Comment]: Unlike the short treatment times required with RF low pressure plasmas, much longer treatments were required when using DBD plasmas to enhance the germination rate of Quinoa seeds.

Task 1 Read the following texts and match feature a–e with the sentence numbers.

Results 1

① The sensory mean values of the biscuits produced from composite flours of wheat and potato flours are shown in Table 4. ② Color and appearance of the biscuits produced with supplementation of potato flour didn't show significant difference ($p < 0.05$) between the control and the 10% supplementation, but there were significant differences between the control and all the other treatments, and there were no significant differences among the treatments used for the biscuit produced. ③ The flavors of the biscuits produced showed significant differences ($p < 0.05$) between the control and all the other treatments, but there were no significant differences among the treatments. ④ All the values obtained in each treatment were in an acceptable range, though the decrement trends were observed. ⑤ The texture of the biscuits produced was in an acceptable range, and there were significant differences ($p < 0.05$) among some of the treatments. ⑥ The highest and lowest were recorded in 0 and 50% supplementation level of potato flour with values of 8.22 and 6.5 respectively on a 9-point hedonic scale. ⑦ The taste of the biscuits produced with each supplementation of potato flour was in acceptable range with the highest and lowest recorded in 0 and 50% with values of 8.02 and 7.12. ⑧ There were significant differences ($p < 0.05$) among some of the treatments. ⑨ The overall acceptability of the biscuits produced from wheat and potato composite flours showed significant differences between the control and the other treatments. ⑩ But there were no significant differences between each treatment. ⑪ In general, the produced biscuits were in an acceptable range. ⑫ That is, all the biscuits produced were acceptable. ⑬ Therefore, it is promising that the partial substitution of potato flour for biscuit production enhances the nutritional values and reduces the cost expended on wheat importations.

a. Research question ________ b. Location statement ________

c. Comments ________ d. Specific data ________

e. Major results ________

Results 2

① The majority of RCFs decreased with increasing biochar application rates (Figure 2). ② For example, the RCFs of sulfadiazine and sulfamethoxazole at the 1 mg/kg pharmaceutical treatment were decreased from 3.66 g/g and 2.45 g/g in the unamended soil to 1.35 g/g and 0.99 g/g in the soil amended with 0.1% of biochar, and then to 0.18 g/g and 0.23 g/g in the soil amended with 1% of biochar, respectively. ③ Specifically, at 0.1% of biochar amendment (Figure 2): (i) The RCFs of sulfadiazine, sulfamethoxazole and lincomycin were significantly reduced ($p < 0.05$) by 32.9%–63.1% compared to the control at both pharmaceutical treatment levels. (ii) On the other hand, the RCFs of caffeine, carbamazepine, lamotrigine, carbadox, trimethoprim, oxytetracycline, monensin and triclosan did not significantly change ($p > 0.05$). (iii) The RCFs of tylosin and estrone were significantly reduced ($p < 0.05$) by 36.7%–39.0% at the 1 mg/kg pharmaceutical treatment, whereas both chemicals were not detected in the radishes grown in the biochar amendment soils at the 0.1 mg/kg pharmaceutical treatment. (iv) The RCF of acetaminophen was significantly reduced ($p < 0.05$) by 35.3% at the 1 mg/kg pharmaceutical treatment, but not changed ($p > 0.05$) at the 0.1 mg/kg pharmaceutical treatment. At 1% of biochar amendment, the RCFs of all the 14 measured pharmaceuticals were significantly reduced ($p < 0.05$) by 28.9%–95.0% at both pharmaceutical treatment levels (Figure 2). ④ Compared with the unamended soil, two sulfonamides experienced the greatest inhibition in the plant bioaccumulation for 1% of biochar amendment with 86.7%–95.0% reduction, followed by lincomycin (74.5%–85.8%), carbamazepine (78.9%–82.5%).

a. Research question ________ b. Location statement ________

c. Comments ________ d. Specific data ________

e. Major results ________

2. Sentence structures frequently used in Results

In each element of a Results section, there are some sentence structures frequently used, which are helpful in writing a Results section or identifying the elements of a Results section.

Elements	Sentence structures
Research questions	In this paper, we propose/hypothesize ... In an effort to tackle this problem of ..., we have developed ... The experiment to be reported here was ... The paper proposes/explores/discusses ...
Location statement	As shown/illustrated/presented/given can be seen in Table 1 / Fig. 3 ... As has been proved/demonstrated in Table 2 / Fig. 3 ... The table below illustrates ... The pie chart above shows that ... It can be seen from the data in Table 1 ...
Major results	The results of this study show/indicate/support that ... The current study has found/confirmed that ... We observe a direct relation between ... and ... From the investigation of ..., it is found that ... Overall, these results indicate/reveal that ... Taken together, these results suggest that there is an association between ... In summary, these results show/suggest that ...
Specific data	There was no increase of ... A positive correlation was found between ... Interestingly, there were also differences in the ratio of ... The most striking result to emerge from the data is that ... The percentage/number of ... has almost doubled as compared with ... There was a ... percent increase/decline in the number/rate ... It accounts for / makes up / constitutes ... percent of the total. The number jumped/incensed/decreased/fell/rose by ... percent to ... percent.
Comment	This also accords with our earlier findings. In contrast to earlier findings, however, no evidence of ... was detected. These results are inconsistent with prior research indicating that ... There are two likely causes for the differences between ... It is difficult to explain this result, but it might be related to ... The phenomenon/trend/results could contribute/lead to ...

Task 2 **Read the above two Results in Task 1 again and find out the signal words.**

	Research questions	Location statement	Major findings	Specific data	Comment
Results 1					
Results 2					

5B Organization: *General-specific (GS)*

The general-specific text starts with a generalization. Following the generalization, specific information is provided using logical connectors. A Results section usually follows the GS pattern. It usually begins with a major result and a location statement to present the general findings. And then, it goes on to illustrate the key points with specific data to support the major result. So, in terms of organization, it features the GS pattern.

Example 1

① The results indicated that the effects of irrigation and soil tillage-sowing practices on CO_2 emission were significant (Figure 1 and Table 1). ② In both years, full irrigation with wastewater and conventional tillage had the highest CO_2 emissions (Figure 1). ③ While the CO_2 emissions in the second year were found higher than the first year values in all irrigation treatments in the conventional tillage practice, the CO_2 emissions higher than the first year were determined under fully irrigated conditions in the direct sowing practice. ④ It could be said that the high organic carbon contents in the second year can appear in this finding. ⑤ However, deficit irrigation treatments in direct sowing conditions probably resulted in fewer CO_2 emissions in the second year due to a change in the organic matter dynamics by affecting soil biology (Li et al. 2010). ⑥ On a 2-year average, the CO_2 emission in the WW100 treatment (0.263 gm-2h-1) was found to be 23.4%, 25.0%, and 59.3% higher, respectively, compared to FW100, WW67, and WW33 treatments. ⑦ Direct sowing (0.193 gm-2h-1) practice resulted in 17.0% fewer CO_2 emissions than the conventional tillage.

In the result, we can see the GS pattern:

Sentence ① states the major result — the effects of irrigation and soil tillage-sowing practices on CO_2 emission were significant.

Sentences ②–⑦ present the specific data to illustrate the relationship between irrigation, tillage-sowing and CO_2 emission.

Example 2

① There were no significant differences in phenolic compound content or antioxidant activity between 80% methanol and 80% ethanol. ② The phenolic compound content and antioxidant activity were highest in the 80% acetone extraction solution, followed by those in 80% methanol and 80% ethanol extraction, and the lowest in pure water extraction. ③ A similar result was also reported in the extraction of phenolic compounds from lychee pulp. ④ However, Zhao et al. found that the 80% methanol extract showed higher O_2, DPPH, and ABTS+ scavenging activities than the 80% ethanol. ⑤ Therefore, the methanol and acetone solvents were used to extract HBP in subsequent experiments in this study.

In the result, we can see the GS pattern:

Sentence ① states the major result — no significant differences in phenolic compound content or antioxidant activity.

Sentences ②–④ present the specific data to illustrate the major result.

Sentence ⑤ responds to the result — the methanol and acetone solvents were used to extract HBP.

Task 1 Read the following text and finish the exercises.

① Research has clearly shown that it is not the total concentration but the reactive fraction of heavy metals in soil that dictates their toxicity to plants, microbes and human beings. ② The water-soluble and exchangeable forms of heavy metals are much more reactive and bio-available than the precipitated species (Roberts et al., 2005; Leitgib et al., 2007; Kim et al., 2015). ③ Distribution of the heavy metal species, however, is influenced by a number of environmental factors, particularly the soil pH, redox potential, clay content and presence of other cations and anions in the soil solution (Rieuwerts et al., 1998; Pietrzykowski et al., 2014). ④ To precisely evaluate the bioavailability of heavy metals in soil, these influencing factors have to be considered.

1) This Results opens with:
   A definition ☐
   Some data ☐
   A major result ☐
2) This Results is organized in the pattern of:
   General-specific ☐
   Specific-general ☐
3) Finish the organization of the Results:

Items	Sentence numbers
Major result	
Specific details	
Response to the result	

Task 2 The following are sentences from a Results section in GS structure. Restructure them by writing the numbers in the correct order.

① However, additional analyses showed that phthalic acid, benzoic acid and lauric acid induce the swarming motility of R. solanacearum in a dose-dependent manner, and no significant difference was found between these three organic acids and the control treatment at some concentrations (Fig. 4).

② A swarming assay was performed to assess the effects of these organic acids on the swarming motility of *R. solanacearum* on 0.35 % plate agar.

③ These results suggest that different organic acids differentially induce the swarming ability of *R. solanacearum*.

④ At various concentrations, myristic acid exerted marked effects on swarming motility, and the motility diameters obtained at concentrations ranging from 50 to 200 μM reached nearly 4.22-, 4.25-, 4.37- and 3.89-fold higher values than those of the control, respectively.

⑤ Furthermore, cinnamic and fumaric acids also markedly stimulated motility compared with the other treatments.

⑥ The results showed that all six organic acids at concentrations ranging from 50 to 200 μM induced swarming motility (Fig. 4).

Correct order: ________________

5C Rhetorical Function: *Graph Description*

The data in the Results section are usually presented in graphs which are always accompanied by explanatory text to highlight and interpret significant facts. A graph is an illustrated representation of data that helps people understand the information. Graphs are more useful for presenting data over a period of time. Commonly used in academic papers, graphs usually show how two or more sets of numbers or measurements are related to each other. The function of a line graph is to describe a TREND. You should therefore try to describe the trend in it. In other words, if there are many lines in the graph(s), you need to give a general description of the trend. If there is only one or two, then you can go into more detail. So, when you describe the movement of the line(s) in the graph, you need to give a numerical detailed description at the important points of the line.

There are some words or phrases which are frequently used in graph description.

1. Description about the trend in graphs

1) Verbs or verb phrases about trends in graphs

Rising trend: rise, go up, increase, grow, climb, ascend, jump, surge, soar, shoot up, rocket, boom, double, triple, gain, leap, accelerate, peak, mount spike, balloon, mushroom, recover, improve

Falling trend: fall, sink, drop, dip, reduce, go down, decrease, descend, slide, slip, decline, reduce, shrink, dwindle, be halved, tumble, plummet, plunge, collapse, slump, diminish

Even trend: stabilize at, steady, show no/little change, remain stable (at), remain steady (at), stay (at), stay constant (at), maintain the same level, level off (at)

Fluctuate: fluctuate, rise and fall, go up and down

Comparison: as well as, not so much ... as ..., more A than B, more than ..., larger than, higher than

Examples

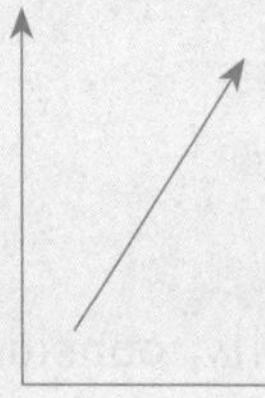
①
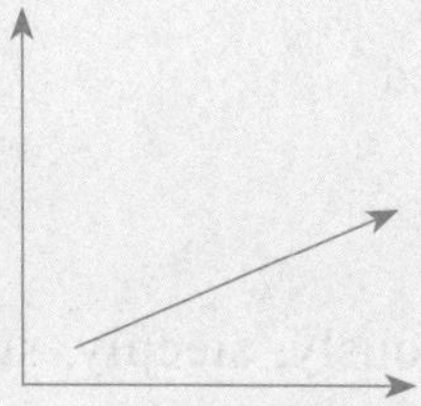
②
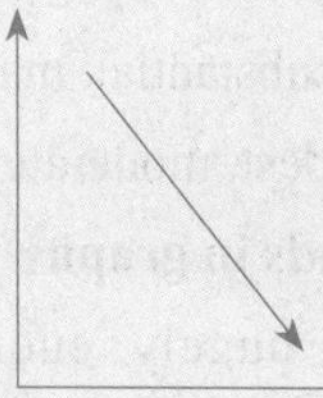
③
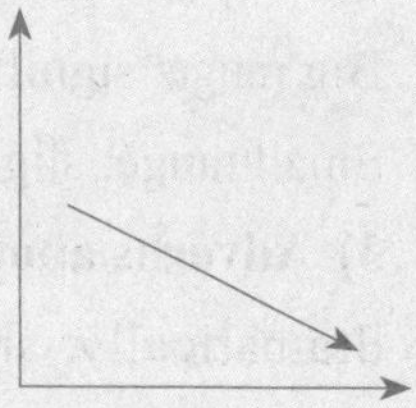
④

① rise dramatically/sharply
② increase steadily
③ drop sharply
④ fall gradually

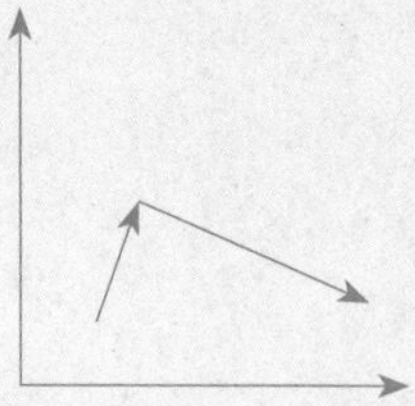
⑤
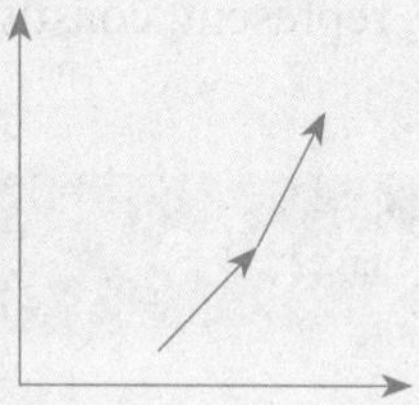
⑥
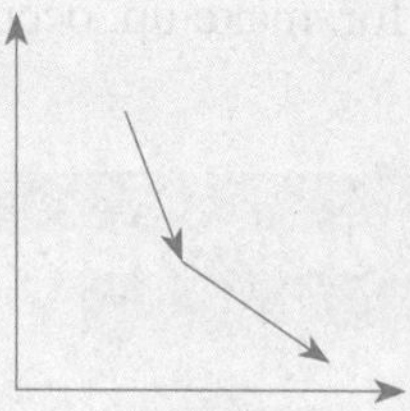
⑦
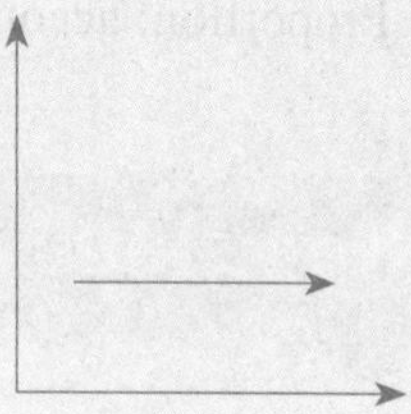
⑧

⑤ The sharp increase is replaced by a slight drop.
⑥ The increase sped up.
⑦ The decrease slowed down.
⑧ remain steady / level off

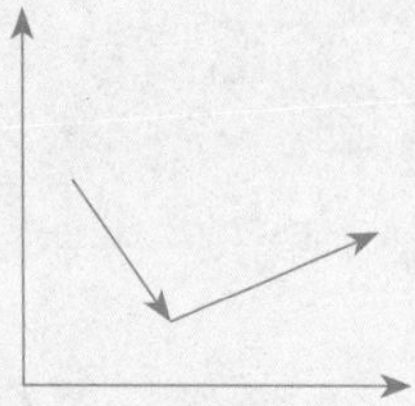
⑨
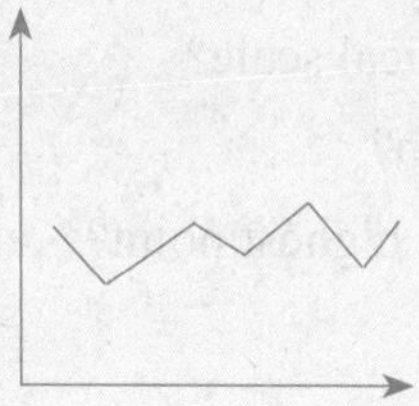
⑩
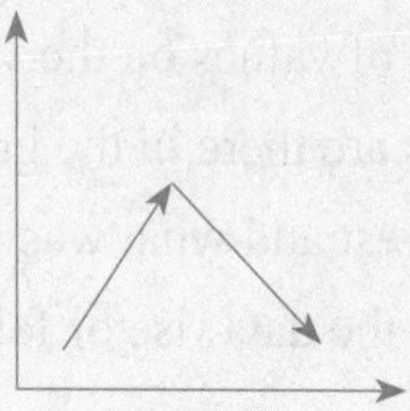
⑪
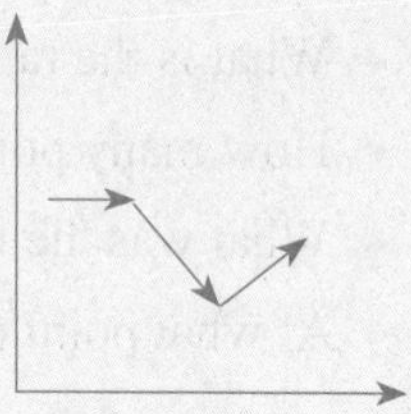
⑫

⑨ There is a slight increase after the dramatic fall.
⑩ fluctuate
⑪ peak
⑫ recover

2) Adjectives about trends in graphs

Fast speed: sharp, rapid, dramatic, drastic

Slow speed: gradual, slow, steady, gentle

Big range: significant, substantial, marked

Small range: slight, modest, moderate

3) Adverbs about trends in graphs

dramatically, sharply, hugely, enormously, steeply, substantially, considerably, significantly, markedly, moderately, slightly, minimally, rapidly, quickly, swiftly, suddenly, steadily, gradually, slowly, heavily

4) Other phrases about trends in graphs

The highest point: peak at, reach the highest point at

The lowest point: bottom out at, reach the bottom at, the lowest point at

The point arrived: reach, arrive at, amount to, hit, stand at

Proportion: account for, make up, occupy, represent, constitute

2. Key skills of describing graphs

1) Asking and answering questions about graphs

Graph should definitely be described in relation to specific topic. To write the description well, perhaps you can start with questions to get familiar with the graph. You could ask yourself the following questions:

- What is the title of this line graph?
- What is the range of values on the horizontal scale?
- What is the range of values on the vertical scale?
- How many points are there in the graph?
- What was the lowest and what was the highest point?
- At what point did the data rise or fall?

Example

For Figure 1, the following questions could be raised:

1. What was the line graph about?

2. What was the busiest time of day in the store of agricultural products?

3. What was the largest number of people in the store?
4. What was the lowest number of people in the store?
5. At what time does business start to slow down?

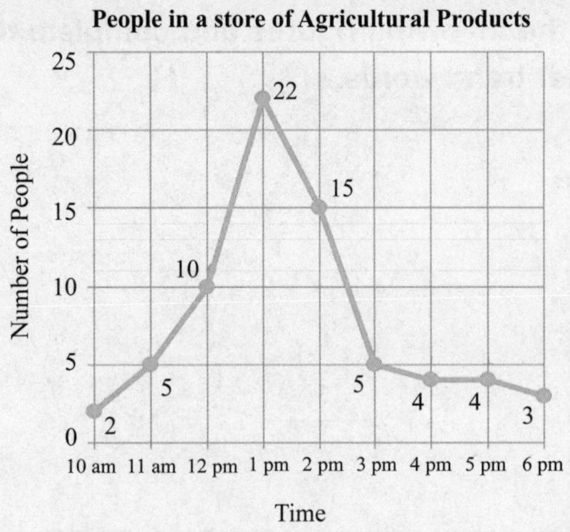

Figure 1 People in a Store of Agricultural Products

2) Making notes

After reading the graph through Q&A, you may combine the trends with the topic and make notes of what you are going to say.

Study Figure 1, and try to make some short notes.

Example

1. From 10 am to 1 pm, the number of people in the store increased significantly.
2. The number of people peaked at 1 pm at approximately 22.
3. Between 1 pm and 3 pm, the number of people decreased sharply.
4. Between 3 pm and 6 pm, there was a slight decrease.

3) Other tips

If you are weak in English grammar, try to use short sentences. This will make it easier for you to control the grammar and the meaning of your writing and will contribute to a better cohesion and coherence mark. Besides, think about the tenses of your verbs. If you're writing about something that happened in the past, your verbs will need to be in the past tenses. If you're describing the future, you will need to use the future tenses. If it's a habitual action, you'll need the present simple tense.

Task **Look at the following figures and complete the information using the proper trend words.**

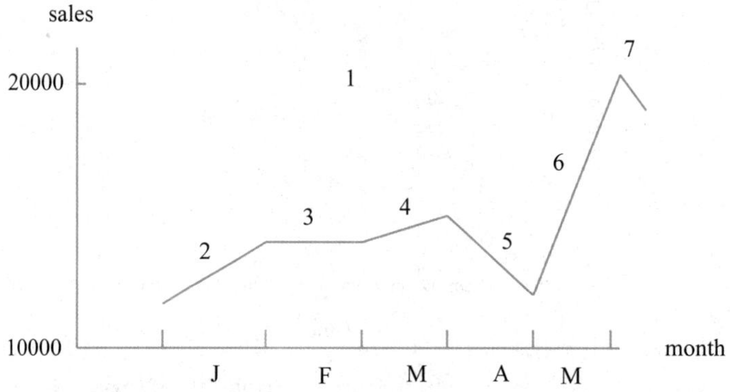

Figure 1

1) As the graph shows, there was an ________ in sales of agricultural products.
2) In January, sales ________.
3) In February, sales ________.
4) In March, sales ________.
5) There was a ________ in April, then a ______ to a ________ of 20,000 in May.

Figure 2

After a sharp ________ in 1996, orders of agricultural products in F city ________ for 12 months and then ________ again in 1998.

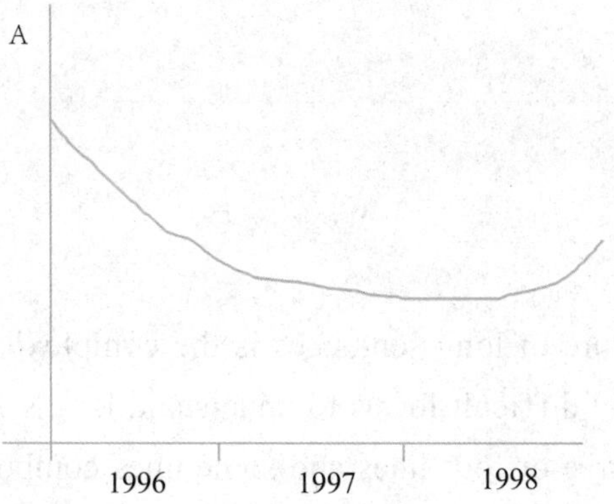

Figure 3

After ________ steadily for two years, orders of agricultural products in A city finally ________ and began a ________ in 1998.

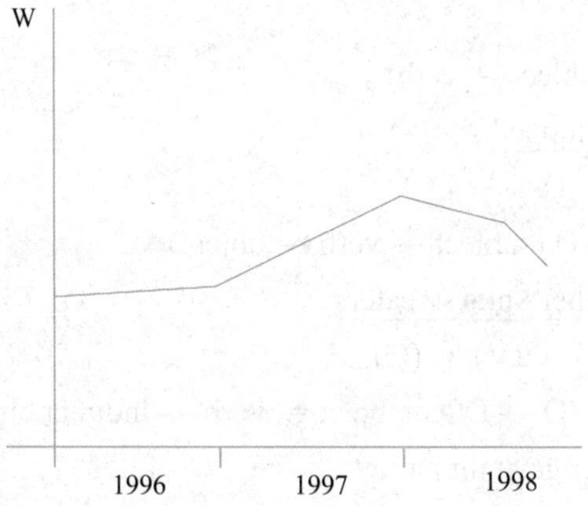

Figure 4

Orders of agricultural products in W city ________ sharply in 1997 but ________ at the end of the year and then ________ back to their 1996 levels.

5D Register: *Long Sentences*

The distinctive feature of long sentences is the complexity of sentence composition and structure. It is usually difficult for us to understand long sentences at a glance. A long sentence usually takes three or four lines and sometimes composes a paragraph. The most important step in analyzing long sentences is to identify the sentence patterns. There are the five basic sentence patterns in English.

1. Basic sentence pattern

Pattern 1: S — V (subject — verb)

The rain stopped.
(S) (V)

Pattern 2: S — V — O (subject — verb — object)

Grandmother knits sweaters.
(S) (V) (O)

Pattern 3: S — V — IO — DO (subject — verb — indirect object — direct object)

His mother gave him a toy.
(S) (V) (IO) (DO)

Pattern 4: S — V — P (subject — verb — predicative)

You are handsome.
(S) (V) (P)

Pattern 5: S — V — O — C (subject — verb — object — objective complement)

We find it interesting.
(S) (V) (O) (C)

In academic articles, long sentences enable authors to make what they write seem serious and formal, which is mainly achieved by complex grammatical structures of the language. Such long sentences cover much more information than simple ones. Subordinate clauses and absolute phrases are common and prevailing devices for long sentences. Here are three main types of long sentences.

2. Types of long sentences

1) Simple sentences with modified and complimented components

A simple sentence consists of an independent clause, which means it contains a subject and a verb. And a simple sentence can have a compound subject, a compound verb or both. In the following example, the subject and the verb are underlined.

e.g. *My brother and I were taught but not licensed to fly a plane.*

(S) (V)

However, in order to enrich the sentences and make the meaning concise and dense, there are many components to modify and complement the simple sentence. For example, non-finite verbs, the nominative absolute, the long pronominal and postnominal modifiers, and insertions are usually included in the long sentence. In the following example, the modified and complimented components are put in the parentheses and the sentence trunk is underlined.

e.g. *(Having chosen family television programs and women's magazines), the toothpaste marketer, (for instance,) must select the (exact television) programs and stations as well as the (specific women's) magazines (to be used).*

In this simple sentence of more than 30 words, the *v.*-ing form phrase is used as an adverbial of time, and the infinitive "to be used" modifies not only "women's magazine", but also "television programs and stations". "Exact television" modifies both "programs and stations". The insertion "for instance" separates the subject from the predicate.

2) A compound sentence

A compound sentence consists of two or more independent clauses. Independent clauses should be combined in a compound sentence only if they are closely related.

e.g. *The highest weight was recorded in control (100% wheat) and the lowest was recorded*

(Independent clause 1)

with the supplementation of 50% potato with values of 8.26 g and 7.61 g respectively.

(Independent clause 2)

3) A complex sentence

A complex sentence consists of an independent clause and one or more subordinate clauses. A subordinate clause, like a phrase, can be used as an adverb, an adjective, or a noun in a sentence. However, a clause has both a subject and a verb, but a phrase does not.

e.g. I can't live in fear of the possibility that as the earth's population grows and we use more and more of our non-renewable (不能再生的) resources, our children may have to lead poorer lives.

The apposition clause is guided by "that" after "the possibility". And it also contains a temporal adverbial clause with a parallel structure "as".

e.g. The result of the physical analysis of the biscuits produced from wheat and potato flour blends is shown in Table 1, which shows that the supplementation of various levels of composite flours has a significant effect on weight, diameter and thickness.

The non-restrictive attributive clause is guided by "which", and it contains an object clause led by "that".

4) A compound-complex sentence

A compound-complex sentence consists of two or more independent clauses and one or more subordinate clauses.

e.g. The protein decreases as the supplementation level of potato increases from 0 to 50%, but there were no differences among the treatments.

The two individual clauses was connected with "but", and the first individual clauses contains a temporal adverbial clause guided by "as".

3. Ways to analyze the long sentences

When analyzing a long sentence, first of all, don't be intimidated by the fact that the sentence is too long, because no matter how complex a sentence is, it is composed of some basic components. Secondly, it is necessary to identify the syntactic structure of the English text, find out the central content of the whole sentence and its various layers of meaning, and then analyze the mutual logical relationship between the different layers of meaning. The main steps of analyzing long sentences are as follows:

(1) Identify the sentence patterns. Identify the subject, predicate and object of the whole sentence, and grasp the structure of the sentence as a whole.

(2) Find out all the predicate structures, non-predicate verbs, prepositional phrases and leading words of clauses in the sentence.

(3) Analyze the function of clauses and phrases, for example, whether they are subject clauses, object clauses, or predicative clauses.

(4) Analyze the relationship between words, phrases and clauses, for example, which antecedent is modified by the attributive clause, etc.

(5) Pay attention to other components such as insertions.

(6) Analyze whether there are fixed phrases or fixed collocations in the sentence.

Task 1 **Label each sentence as simple, compound, complex, or compound-complex, underline the basic sentence pattern with the straight line and the guide words with the wave, and circle the conjunctions if necessary.**

> **Model:** The processing of potato into flour is perhaps the most satisfactory method of creating a product that is not only functionally adequate, but can also be kept for an extended period without spoilage. (complex)
>
> Phenolic acids were the most abundant class of phenolic compounds and they accounted for 89% of the amount of phenols identified in the UG extract. (compound)

1) The texture of the biscuits produced was in acceptably range, and there were significant differences ($p < 0.05$) among some of the treatments. (   )
2) The fiber content of the biscuits produced increases as the supplementation level of potato flour increases, with the highest and lowest value being 6.80 g/100 g and 4.64 g/100 g at 50% and 0% substitution of potato flour respectively. (   )
3) This work is in line with the work of Sneha et al. (2012), in which they reported as the supplementation of sweet potato flour increases for the production of biscuits, the fiber content of the final product also increases. (   )
4) Therefore, research toward the development of wheat biscuit products will address the challenges of nutrition deficiencies and post-harvest losses of potatoes. (   )
5) Okaka (1997) described the production of biscuits as a mixture of flour and water, but he also indicated that it may contain fat, sugar and other ingredients mixed together into a dough which is rested for a period and then passed between rollers to make a sheet. (   )
6) An approach in the present study is to replace the wheat flour in biscuits with potato flour (gluten-free flours) in order to increase the fiber and other nutrients. (   )

4. Ways to write long sentences

Since long sentences are more serious and solemn, it is necessary to use long sentences in academic papers. Usually, the sentences are enriched by adding modifiers, adverbials, insertions, subordinate clauses, non-finite verbs and various phrases.

Task 2 **Rewrite the short sentences using the devices of long sentences.**

> **Model:** Short sentence: Agriculture is the sector.
> Long sentence: Agriculture is the sector that consumes the most fresh water resources compared to other sectors.

(1) Short sentence: Factories discharged gas and liquid.
Long sentence: ______________________________

____

____

(2) Short sentence: Our results demonstrated the cause.
Long sentence: ______________________________

____

____

(3) Short sentence: The intestinal system is crucial.
Long sentence: ______________________________

____

____

(4) Short sentence: Soil-borne disease is a major obstacle.
Long sentence: ______________________________

____

____

(5) Short sentence: The temperature was monitored.
Long sentence: ______________________________

____

____

5E Cohesion (4): *Substitution*

Substitution is a grammatical device for avoiding repetition and achieving textual cohesion. Generally, there are three categories of substitution: nominal substitution, verbal substitution and clausal substitution.

1. Nominal substitution

Nominal substitution means the replacement of a noun or a noun phrase with a nominal substitute. For example, the words *one/ones*, *the same*, *the kind*, *the sort*, and some indefinite pronouns, such as *all*, *both*, *some*, *any*, *enough*, *several*, *none*, *many*, *much*, *(a) few*, *(a) little*, *the other*, *others*, *another*, *either*, *neither*, etc.

1) one, ones

The substitute "one/ones" is usually used as the subject of a noun phrase and can only be used to replace a word that is itself the subject of a noun phrase.

① "One" substitutes a singular count noun or phrase, and "ones" substitutes a plural noun or phrase.

② "One" could be used alone or with a premodifier or postmodifier; whereas "ones" have to be used with a premodier or postmodifier.

Examples

1. There are some apples. You may take one. (one = an apple)
2. I haven't got any stamps. Could you give me one? (one = a stamp)
3. These apples aren't ripe. Give me some / the ripe ones. (ones = apples)
4. I wish I'd bought a few jars of honey. Did you notice the ones they were selling by the roadside? (ones = jars)

5. We are apt to find that the very men who block a scheme are the ones who clamor the loudest. (ones = men)

③ indefinite article / possessive determiner + premodifier + one/ones

Example

Let's take your/Jack's new one/ones.

④ definite article + one/ones + postmodifier

Examples

1. The one/ones in the corner is/are mine.
2. This room and the one upstairs are well furnished.

2) that, those

"That" can substitute singular uncountable noun or nominal phrase and can only be used for non-personal reference. "Those" can substitute plural nouns, including the coordination construction "with"/"and". It can be used for personal as well as non-personal reference.

Examples

1. The poison of the cobra is more deadly than that of the rattle snake. (that = the poison)
2. The blonde girls I saw were older than those you were dancing with. (those = the girls)
3. His speech and behavior were those of a hooligan. (those = speech and behavior)

3) Similarity and disparity between one/ones and that/those

Both can be used to replace a count noun or a countable noun phrase, but if "of" phrase follows, we can only use "that" rather than "one"; when an attributive clause follows, we will use "one" more often.

Examples

1. My seat was next to that of the mayor. (that = seat)
2. It was a marvelous picture just like the one (that) we had seen the other day. (one = picture)
3. I drew my chair nearer to the one Sophy was sitting on. (one = chair)

4) Other words

As mentioned above, there are other nominal substitutions such as *some*, *any*, *none*, *many*, *much*, *enough*, *each*, *(n)either*, *both*, *all*, etc.

Examples

1. Two boys entered, each carrying a suitcase.
2. There are two roads into town, and you can take either.
3. Proust and James are great novelists, but I like Tolstoy better than either.
4. She and her husband went to see the house. Both felt it suitable.
5. The soup is delicious, and the turkey smells the same.

2. Verbal substitution

Verbal substitution is the replacement of a verb element with a verbal substitute — "do" or "do so". The verb "do" can replace a main verb or a verb plus complement.

Examples

1. A: Does he promise to come tonight?

B: Yes, he does.
2. I prophesied he would fail, and he did.
3. She plays the piano better than she does the guitar.
4. The bomb doesn't change our lives, but the computer does.

However, substitute "do" can not replace "have", "be" and non-finite verbs.

Examples

1. I am not as clever as he is. (We don't say "I am not as clever as he does".)
2. Your mother has no right, not any more than I have, to tell you how to run your life. (We don't say "Your mother has no right, not any more than I do".)
3. A: Peter hunts rabbits.
   B: Yes, he wanted me to hunt rabbits, too. (We don't say "yes, he wanted me to do, too".)

3. Clausal substitution

1) The words "so" and "not" could commonly be used to replace that-clauses that represent a belief, an assumption or an emotion with a tone of uncertainty or tentativeness. These substitutes normally go with the following words such as appear, believe, expect, fancy, fear, gather, guess, hope, imagine, presume, reckon, seem, suppose, suspect, think, trust, etc. If "so" or "not" goes with "appear" or "seem", we will always use "it" at the beginning of the sentence.

Examples

1. A: Do you think he'll come tomorrow?
   B: Yes, I think so. / No, I think not.

2. Many people believe that the international situation will deteriorate. My father thinks so, but I believe not.
3. A: Is there going to be a holiday tomorrow?

   B: I believe so.
4. A: Is it true that Geoff has had a heart attack?

   B: It seems so.

2) As clausal substitutes, "so" and "not" can go with "if" to form verbless clauses "if so" and "if not".

Examples

1. If it clears, we'll go out; if not (= if it doesn't clear), not.
2. He may be busy. If so (= If he is busy), I'll call later. If not (= If he is not busy), can I see him now?

3) "So" and "not" can go with "afraid"

Examples

A: Has the news reached home yet?

B: I'm afraid so/not.

Task 1 Improve the following sentences with substitutions.

1) There are good books about veterinarian as well as bad books about veterinarian.
2) The experimental facility didn't work, so we want to buy a new facility.
3) They sent us some small apples as experimental subjects instead of big apples.
4) They don't buy their agricultural foods at the local supermarket, but we buy our foods at the local supermarket.
5) Pam smokes a lot. Does his brother smoke a lot?

Task 2 Identify the cohesive ties in the following text and write them in the corresponding box.

Title: *Sweetpotato Cultivars Differ in Efficiency of Wound Healing*

Result

The effectiveness of wound healing in protecting the wound against pathogen invasion was tested by placing the mycelia of Rhizopus oryzae directly onto wounds at various stages during the healing process. Fig. 4 shows the dimensions of lesions allowed to develop over 2 days for 12 cultivars. SPK004 is notable in the development of lesions on wounds even after 6 days of healing. Analysis using contingency tables relates the incidence of rots either to the presence of lignin, or to the completeness of the lignified layer. Wounds with a complete lignified layer, i.e. a lignification score of 1, rotted less frequently than those with just the presence of lignin, i.e. lignification scores greater than 0. This was highly significant at 3 days ($p = 0.010$, Fisher's exact test $p = 0.0154$) and 6 days after wounding ($p < 0.001$). It appears that pathogens are unable to effectively degrade mature suberin (Kolattukudy, 1981, 1984), and the same is likely to be true for ligno-suberin complexes.

Cohesive ties					
Grammatical cohesion			Lexical cohesion		
Conjunction	Substitution	Reference	Lexical reiteration: repetition, synonym, superordinate	Lexical chains	General nouns
			╳	╳	

Task 3 Fill in the blanks with the substitutions given in the following box.

this	that	one	neither	such

Where is capital as input? Essentially, it is treated as part of (intermediate) industrial inputs for the following two reasons. First, the role of fixed capital in agriculture is not as important as 1) ________ in industry, because in China, farming remains, in most areas, largely small in scale and labor intensive. Probably the most important piece of fixed capital used in agriculture is tractors. However, in 2) ________ case, fuel and maintenance costs are arguably much more important than depreciation. In practice, it is rather difficult to separate the cost of fixed capital from intermediate inputs in China's agricultural production. Second, even if we want to include capital as a separate input, there is no reliable data on the quantity of aggregate fixed capital used in agriculture. 3) ________ the share of costs nor the quantity of agricultural fixed capital is directly measured. As a matter of fact, net agricultural output does not include depreciation as an element of value added. Previous studies used various proxies as measures of agricultural capital, such as draft animals and horsepower of agricultural machines. Although 4) ________ proxies may be appropriate for a microeconomic study, it is doubtful that they are suitable for studies using aggregate data such as the present 5) ________.

5F Assignments: *Reading Comprehension and Vocabulary*

Part one Reading comprehension

Task 1 Read the following text and tell whether the statements that follow are true (T) or false (F).

metabolic *adj.* 新陈代谢的
R. solanacearum *n.* 青枯
cinnamic *adj.* 肉桂的
myristic *adj.* 肉豆蔻的
incubation *n.* 潜伏期
retention *n.* 保留

To investigate the **metabolic** response of ***R. solanacearum*** to organics, the persistent levels of **cinnamic** and **myristic** acid in M63 medium were determined by HPLC and GC. The results showed that the levels of cinnamic and myristic acids changed only slightly throughout the rest of the **incubation** period. The **retention** rates of cinnamic and myristic acids were 95.82% and 93.42% at 12 hpi and 90.22% and 88.43% at 24 hpi, respectively (Fig. 2). These results confirm that these organic acids cannot serve as carbon sources for *R. solanacearum*.

______ 1) This Results section has all the elements of a Results section.
______ 2) This Results section starts with a major result.
______ 3) There are no substitutions in this Results section.
______ 4) There are specific data in this Results section.
______ 5) Past tense is the only tense used in this Results section.
______ 6) This Results section is organized in general-specific pattern.

Task 2 Read the following text and answer the questions that follow.

chronophotograph *n.* 连续照片

① As shown by the **chronophotographs** in panels (a)–(b), the simulations agree qualitatively well with experiments for both events. ② In addition, the time evolution of the settling speed, reported in panels (c)–(d), shows excellent quantitative agreement, thus confirming the **adequacy** of our method to capture fiber-obstacle interactions in a **viscous** fluid. ③ In the gliding case, the fiber starts close to its **equilibrium** shape with a settling speed larger than a rigid one (V > V1). (See panels (a) and (c)) ④ As it settles, it **progressively** slows down until it partially hits the obstacle and reaches its minimum speed. ⑤ The speed increases again as the fiber glides along the obstacle. ⑥ Upon release, it is more **aligned** with gravity and reaches its maximum **velocity** (V/Vi ~ 1.2). ⑦ Eventually, it slowly relaxes to its equilibrium shape: the velocity decreases to its initial value and the fiber drifts sideways as it **reorients** perpendicular to gravity. ⑧ In the trapping case, the initial lateral distance between the fiber and obstacle, Dx, is smaller. ⑨ As a result, the fiber slows down continuously, wraps around the obstacle and finally remains in a trapped **configuration** indefinitely. (See Fig. 3(b) and Fig. 3 (d)) ⑩ The decrease in the ettling speed is more pronounced when the fiber touches and wraps around the obstacle in the **interval** 1.3 < t/T < 4, where T = Ln/W is the characteristic settling time. ⑪ The trapped configuration in **the steady regime** is **asymmetric** with respect to the obstacle center of mass. ⑫ Such asymmetry originates from the **tangential** contact forces and results from an equilibrium configuration that minimizes the total energy due to **external** (gravity and contact) and **internal elastic** forces.

adequacy *n.* 适当，恰当
viscous *adj.* 黏性的
equilibrium *n.* 平衡
progressively *adv.* 逐渐地
align *v.* 使对齐
velocity *n.* 速度
reorient *v.* 重新定向，再调整
configuration *n.* 配置，布局
interval *n.* 间隔
the steady regime 稳定状态
asymmetric *adj.* 不对称的
tangential *adj.* 正切的
external *adj.* 外部的
internal *adj.* 内部的
elastic *adj.* 弹性的

1) How many of elements can you find in this Results section? Identify the sentences that correspond to each element in the table below and find out the substitutions and words describing trends in each one.

Typical Elements	Sentences	Substitutions	Words describing trends
Research questions			
Location statement			

(To be continued)

(Continued)

Typical Elements	Sentences	Substitutions	Words describing trends
Major results			
Specific data			
Comment			

2) Does the author use the first-person pronoun (I or we)? What is the case in your field?
3) Are there any long sentences in the text? If so, please underline and analyze them.

Task 3 Write a Results section for a research you have been involved in.

Part two Vocabulary

A trend is the general direction in which something is developing or changing over time. A projection is a prediction of future change. Trends and projections are usually described in the Results part. Some words or expressions, like *rise*, *fall*, *increase*, *decrease*, *grow*, *change*, etc., are commonly used for writing about trends and projections.

Task 1 Choose appropriate words to complete the following sentences.

1) The result indicated that FL could ________ Cy-induced immunosuppressive activity on the development of immune organs.
   a. establish　b. weaken　c. construct　d. extent
2) The International Food Policy Research Institute (IFPRI) estimates that the economic ________ in 2020 could increase the number of people living in extreme poverty by 20% or 140 million people.
   a. boom　b. principle　c. contraction　d. process
3) Dam operations often result in ________ magnitude, variability, and timing of flow, with subsequent impacts on fish populations and communities.
   a. altered　b. assorted　c. resolved　d. occupied
4) To increase the printing time, the tip was exchanged for a shorter one, which provided the opportunity to increase the printing speed and ________ the printing pressure.
   a. generate　b. decrease　c. implement　d. impose

5) Most of the health benefits of tea come from antioxidants that ________ the formation of plaque along the body's blood vessels.

a. accelerate　　b. promote
c. inhibit　　d. remove

6) The altered Zmorph printer had difficulties extruding the ink in a constant manner due to the circling motion of the stepper motor, resulting in high ________ in weight, thickness, and drug amounts.

a. evolutions　　b. declines
c. impacts　　d. variations

7) Plant biopolymers can act as a physical barrier for the phenol stimuli utilized, thus ________ their interactions with sensory receptors and saliva.

a. hindering　　b. promoting
c. combining　　d. transiting

8) The reason behind the increase in flowers' fresh weight might be the ________ of photosynthesis and maximum accumulation of photosynthates due to IBA application.

a. elimination　　b. allocation　　c. enhancement　　d. integration

9) PGRs (plant growth regulators) play an important role in ________ antioxidant capacity of the plants.

a. focusing　　b. aligning　　c. improving　　d. underlying

10) These peptides are inactive within the sequence of the parent protein and can be ________ following enzymatic hydrolysis.

a. involved　　b. suspended　　c. inspected　　d. released

Task 2 Match words 1)–8) to their meanings a–h.

1) enhance	a. to become worse or decline in quality
2) contract	b. to make uniform or similar
3) degenerate	c. to increase or further improve the good quality, value or status of someone or something
4) manifest	d. to make something happen more slowly or less frequently than normal
5) variation	e. to become (or make something become) less or smaller
6) inhibit	f. to become obvious or noticeable
7) homogenize	g. destroy or get rid of something completely
8) extirpate	h. a change, especially in the amount or level of something

Task 3 **Complete the following passage with words from the word box.**

affected	changes	decline	degenerated	extinction
extirpated	homogenization	increase	loss	manifested

Our results suggest that the construction of major dams have 1) ________ the composition and alpha diversity of fish in the Gan River Basin, although we recognize that other environmental 2) ________ independent of dams have occurred. Previous research has shown that after construction of Wanan Dam, Chinese sturgeon (Acipenser sinensis), Reeves shad (Tenualosa reevesii), and long spiky-head carp (Luciobrama macrocephalus) were 3) ________, and population sizes of Japanese grenadier anchovy (Coilia nasus) and yellow cheek (Elopichthys bambusa) rapidly declined in the Gan River Basin. Total reproductive output of four major Chinese carps rapidly declined from 2.5 billion in the 1960s to 20 million in 2000, and the 12 spawning grounds of four major Chinese carps and three spawning grounds of Reeves shad (Tenualosa reevesii) were 4) ________ and disappeared.

River ecosystems have experienced hydrological disconnection and habitat fragmentation due to dam constructions, resulting in biological 5) ________. For example, flow regimes, water quality, and habitat conditions of Yangtze River Basin are significantly affected by the Three Gorges Dam, which resulted in the 6) ________ of many native and endemic fish species and leading to fish faunal homogenization. Similar effects of dam constructions in the Yellow River Basin have also 7) ________. This study also demonstrated a homogenization trend, with a general 8) ________ in fish taxonomic similarity among regions of the Gan River Basin. Past studies also showed that dams caused habitat fragmentation and 9) ________ in many river ecosystems, resulting in the loss of native and endemic species and the introduction of non-native fish species, leading to a 10) ________ in beta diversity and increase in biological homogenization.

Unit 6 Discussion

Main Contents	Learning Objectives
Subgenre: *Discussion*	★ Understanding the function of a Discussion section. ★ Identifying the structure of a Discussion section.
Organization: *Specific-general*	★ Identifying GS and SG models. ★ Learning about SG models above paragraph level. ★ Understanding a text better by analyzing its organization.
Rhetorical Function: *Cause and Effect; Comparison and Contrast*	★ Learning languages used to explain a point with the cause-effect pattern. ★ Learning languages used to explain a point by comparison and contrast.
Register: *Hedging*	★ Understanding the purpose of hedging. ★ Learning common ways of hedging.
Cohesion (5): *Lexical Reiteration*	★ Practicing using lexical reiteration (repetition, synonym and superordinate) in reading and writing academic texts.
Assignments: *Reading Comprehension and Vocabulary*	★ Reading: Understanding the Discussion section better through in-depth reading. ★ Vocabulary: Identifying evaluative adjectives in Discussion.

6A Subgenre: *Discussion*

The Discussion section usually comes after the Results section in an academic article. This section offers a generalized account of what has been learned in the study. Gradually "zooming out" from the research details, the Discussion presents the major findings by the researchers and may also refer back to some statements made in the Introduction, such as the value of the study. In the Discussion section, you should step back and take a broad look at your findings and your study as a whole (Swales & Feak, 2012).

1. The structure of a Discussion

To better understand the function and features of the Discussion section, we need to pay attention to its difference from the Results. If the Results section deals with facts, then the Discussion section deals with points. Facts are descriptive, while points are interpretive. Therefore, Discussion should be more than summaries, and it should go beyond the results. It involves the authors' opinions in the form of analysis, explanation, judgement and comment. In other words, the primary purpose of the Discussion is to show the relationship among the facts observed in the Results.

The content of the Discussion may vary considerably. The authors have some flexibility in deciding which of the possible points to include and which to highlight. In general, it is intended to consolidate your research area, to point out the limitations and provide suggestions for future research. The following components are usually included in the Discussion.

Table 6.1 The Structure of a Discussion

Move 1	Restating the research questions or hypothesis of your study.	required
Move 2	Summarizing the major findings of your study.	required

(To be continued)

(Continued)

Move 3	Explaining the principles, relationships and generalizations shown by the Results.	required
Move 4	Comparing your study with previous ones.	optional but common
Move 5	Discussing the theoretical and practical implications of your study.	optional but common
Move 6	Indicating the strengths and/or limitations of your study.	optional but common
Move 7	Identifying useful areas for further research.	optional and only common in some areas

It should be noted that many Discussion sections go through the Move 1–3 (or part of it) more than once. Commonly, each cycle takes up a paragraph. Furthermore, the more research questions there are to be discussed, the more this cycling is likely to occur. Such cycling can also occur in the Introduction, but it tends to be less common, especially in shorter RPs (Swales & Feak, 2012).

Example 1[1]

Title of the paper: *Community Food Environment Moderates Association Between Health Care Provider's Advice on Losing Weight and Eating Behaviors*

[Restating research questions] This study examined if the community food environment moderates the relationship between receiving weight loss advice from an HCP and consumption of food and beverages in an OW/OB population. **[Summarizing major findings]** Interaction and stratified analyses revealed that receiving HCP's advice on losing weight was associated with a lower reported consumption of total SSB, soda, and sweetened fruit drinks when participants lived near a small grocery store, or far from a supermarket, LSR, or convenience store. However, when participants lived near supermarkets, LSRs, or convenience stores, there was no association between HCP's advice and reported SSB consumption. We found no association with respect to fruit, vegetable, salad or fast food consumption. These results elucidate the complex role of context (i.e.

1 HCP: health care provider; SSB: sugar-sweetened beverages; OW/OB: overweight and obesity; LSR: limited service restaurant

community food environment) on the effect of HCP's weight loss advice.

[Comparing with previous studies] Most studies analyzing the influence of the community food environment have focused on fruit and vegetable consumption (Dunn et al., 2015; Morland et al., 2002; Robinson et al., 2013; Zenk et al., 2005), with only three examining SSB (Duran et al., 2016; Gustafson et al., 2013; Laska et al., 2010). Laska et al. found similar results to our findings in their sample of adolescents; those having a fast food restaurant or convenience store within 1,600 m of home consumed 25% and 24% more SSB, respectively (Laska et al., 2010). Our result of lower reported SSB consumption in the absence of a supermarket also align with the findings from Gustafson et al. who found that those who shopped frequently at a supermarket had higher odds of consuming SSB (Gustafson et al., 2013).

[Offering explanations] The positive association between supermarkets and SSB consumption may be due to the variety of beverages found in supermarkets. A study in São Paulo, Brazil, found that adults who lived in a census tract with a greater variety of SSB were more likely to consume SSB (Duran et al., 2016). Furthermore, supermarkets have been shown to have lower SSB prices compared to convenience stores (Vilaro and Barnett, 2013).

[Further explanations] The lack of association between HCP's advice and SSB consumption when participants lived near supermarkets, LSRs, or convenience stores suggests that receiving HCP's advice on losing weight may not be powerful enough to overcome environmental cues such as ease of access, in-store marketing, and price discounts.

[Discussing implications] However, modifications to the food environment could theoretically enhance the association between HCP's advice and healthy eating behaviors. Recommended strategies for reducing SSB consumption at the population level include limiting access to SSB, creating a cost differential so that healthy beverages are less expensive than SSB, limiting the marketing of SSB, and including SSB related counselling in routine medical care (Center for Disease Control and Prevention, 2010).

[Indicating strengths] There are several strengths to the current study. Participants were mostly low-income and from racial/ethnic minority groups, populations at higher risk of obesity and obesity-related illnesses. In addition, GIS mapping was used to determine outlet proximity to each participant's home rather than grouping participants by census tract, allowing for more precise measurements of the presence and absence of food outlets.

[Indicating weaknesses] This study does have some limitations. The four cities included in the sample are urban, low-income, and with high minority populations. The results therefore, are likely to be generalizable to similar communities. Because data on advice on losing weight and eating behaviors were self-reported and only collected once, there is a possibility of same-source bias and underreporting of unhealthy eating behaviors (Goris et al., 2000). The specific weight loss advice given by the HCP is also unknown. The observed significant differences in consumption behaviors between those obtaining HCP's advice and those not getting similar advice were modest. In addition, while the classification process for food outlets was robust, it was based on purchased data and there is a possibility that some food outlets were missing from the databases or the information was outdated (Powell et al., 2011). Analyses of the food environment only accounted for one type of food outlet at a time. Lastly, the data were cross-sectional and, therefore, the depicted associations should not be construed as cause-effect relationships.

[Suggesting further research] Additional studies are needed to understand moderating factors that influence HCP's weight loss advice, and possible environmental strategies that assist in HCP's weight loss advice being beneficial.

Example 2[1]

Title of the paper: *Dose-related Changes in Respiration and Enzymatic Activities in Soils Amended with Mobile Platinum and Gold*

[Restating hypothesis] In this study, the hypothesis that the presence of Au and Pt would affect soil respiration differently was generally confirmed. [Summarizing major findings] The soil respiration rates in BGR (acidic), PPN (metal/silt) and MNP (basic) soils largely decreased following the addition

1 BGR: Baren Grounds; PPN: Pinpinio; MNP: Minnipa; FLN: Fox Lane

of Au and Pt. [Offering explanations] The soil respiration rate, an index of soil microbial activity, can decrease significantly in polluted soils as a result of the toxicity associated with the pollutants (Verma et al., 2010). Therefore, the observed decrease in soil respiration rates in this study was likely due to Au and Pt toxicity to soil microbial groups. [Comparing with previous studies] While there are limited reports of Au and Pt effects on soil respiration, such reports, when available, showed that increasing the concentration of Pt (as observed in forest soils) decreased soil respiration (Kalbitz et al., 2008). Significant reductions in soil respiration have been reported in most soils contaminated with heavy metals such as Cu, Zn and Cd (Deng et al., 2015), so the trend observed in this study is not unusual.

[Offering explanations for findings contradictory to the hypothesis (increased soil respiration)] Organic matter can also adsorb some heavy metals, forming stable complexes and rendering such metals non-bioavailable to the soil microbiota (Hamid et al., 2020). Therefore, the organic matter in FLN soil was probably beneficial to soil respiration in that it adsorbed most of the introduced Au while making available ample resources for microbial activities (respiration) (Lai et al., 2013). Increased soil respiration in the presence of heavy metals such as Pt and Au could also occur as a result of a stress response, caused by increased diversion of energy from growth to maintenance energy requirements, different pH, and heavy metal bioavailability.

[Summarizing major findings] The effects of the concentration of Au on soil respiration were variable in Au-amended soils. In some soils, such as MNP and to a certain extent PPN soils, increasing metal concentration did not always lead to a substantial reduction in soil respiration rates, suggesting that a lower contamination threshold was required for detectable impairment of soil functionality. The effects of Pt amendment in the same soil types were different, with stimulation or inhibition of soil respiration largely correlated with increasing metal concentration. [Discussing implications] This demonstrated that soil microbial communities responded differently to Au and Pt contamination, which could reflect the different mobility, adsorption and bioavailability of these heavy metals in different soil types. In summary, the hypothesis that the presence of Au and Pt would affect soil respiration differently was generally confirmed. However, the response was varied with both soil type and metal concentration, highlighting the complexity of the interaction of these heavy metals with soils (Chander and Brookes, 1991; Killham, 1985).

[Summarizing major findings] In acidic BGR soils, all the tested enzymes were inhibited by the addition of Au and Pt (except NAG at 1, 25, 100 mg/kg Pt concentration), suggesting that, under acidic conditions, Au and Pt were toxic to most enzymes. **[Comparing with previous studies]** Previous studies have shown similar findings under acidic conditions, with Cu inhibiting enzyme function (Gupta and Aten, 1993; Romié et al., 2014; Tyler and Mcbride, 1982). **[Summarizing major findings]** Acidic environments can also lead to decreased growth in some microorganisms as more metabolic energy is used for maintenance instead of respiration and enzyme function (Sherameti and Varma, 2010). **[Offering explanations]** This may also be related to the mobility of Pt and Au under acidic conditions.

Task 1 **Read the following segments ①–⑧, match them to the elements a–d, and underline the words that prompted your choices.**

Discussion 1

a. Restating research questions
b. Summarizing major findings
c. Offering explanations
d. Comparing with previous studies

______ ① The present investigation for the first time presents a rapid and reliable method for the selection of suitable rose cultivars to be used in blue flower production studies. More specifically, the innate presence of the biochemical base required for the production of desired colors in rose was identified using agrionfilteration.

______ ② Our results revealed that color change to blue was only achievable in cultivars inherently producing dark pink flowers and cultivars with red, yellow, orange and white flowers did not show blue coloration.

______ ③ This was in agreement with the findings of Brugliera et al. (2013), who demonstrated that color change success was cultivar-specific and therefore, cultivar-specific strategies were required as well.

______ ④ Flower pigmentation is due to the accumulation of plant secondary metabolites such as anthocyanins within the vacuoles of epidermal cells (Tanaka et al. 2005). In the present study, the transient expression of pBIH-35S-Del2 resulted in a phenotypic change from pink to mauve in the epidermal cells of some rose cultivars.

______ ⑤ Qi et al. (2013) previously reported similar results in oriental lily "Sorbonne" by the transient expression of PhF3050H.

______ ⑥ Of the three constructs and 30 host cultivars used in the present study, the multi-gene construct pBIH-35S-Del2 in the cultivar "Purple power" showed the most suitable combination for the transformation purposes.

______ ⑦ In a study by Nakatsuka et al. (2007), in which multi-gene cassettes were used as well, synchronous expression of Torenia fournieri DFR, Torenia fournieri ANS and CcF3′5′H produced more delphinidin than the CcF3′5′H alone. The construct used in this study, i.e. pBIH-35S-Del2, contains CcF3′5′H, DFR and ANS. It appears that the expression of these three genes must have led to the generation of dihydromyricetin (DHM) and the reduction of pelargonidin and cyanidin precursors.

______ ⑧ This was in line with the findings of the previous studies in which a more significant accumulation of delphinidin was observed when the two related genes of F3′5′H and DFR were co-expressed in carnation, rose and petunia (Katsumoto et al. 2007; Qi et al. 2013).

Task 2 **Read the following Discussions and match segments ①–⑫ to elements a–h. Each element can be used more than once.**

Discussion 2

Title of the paper: *Effects of Probiotic Litchi Juice on Immunomodulatory Function and Gut Microbiota in Mice*

① Probiotics have the ability to increase immunomodulatory function and modulate gut microbiota composition (Kim et al., 2020; van Baarlen, Wells, & Kleerebezem, 2013). So they are often added to fruit juices, dairy products or meat products to increase their health benefits. Many active ingredients extracted from litchi pulp can improve the immunoregulatory ability of mice (Huang, Zhang, Liu, Xiao, Liu et al., 2016; Huang, Zhang, Liu, Xiao, et al., 2016). However, there were few studies on the healthy function of probiotic litchi juice. Therefore, the aim of this study was to investigate the effects of fermented litchi juice (FL) on immunity and gut microbiota in mice.

② It was found that after fermentation with *L. casei*, the viable cells and TA were both increased, which was consistent with a previous study (Yu et al., 2013). And the content of TP, TF, and EPS were also increased after fermentation. ③ These results demonstrated that the lactic acid bacteria improved the phytochemical concentrations of the juice during the fermentation (Kaltsa, Papaliaga, Papaioannou, & Kotzekidou, 2015; Kwaw et al., 2018; Zheng et al., 2014). Moreover, the increase in TP, TF, and EPS content of probiotic litchi juice caused the improvement of antioxidant capacity, which was similar to previous studies (Kalaycıoğlu & Erim, 2017; Li et al., 2017; Mahdhi et al., 2017).

④ Cy is an effective immunosuppressive agent affecting the organism's immune system, which induces the imbalance of the organism's immune function (Zhu et al., 2019). Therefore, the immunocompromised animal model constructed by Cy-treatment was chosen to evaluate the effect of FL on the immunomodulatory function. Both spleen and thymus are important immune organs in innate immunity, and are the places of lymphocytes differentiating, maturing, and producing immune response (Meng, Li, Da Jin, Lu, & Huo, 2018). The decrease in immune organ indexes indicates the decline of immune function (Huang, Zhang, Liu, Xiao, Liu et al., 2016). ⑤ Compared with the Cy-treated group, the spleen and thymus indexes in the (Cy + FL)-treated group were higher. The result indicated that FL could weaken Cy-induced immunosuppressive activity on the development of immune organs. ⑥ Park et al. also found that the organism immunity was promoted by increasing spleen and thymus indexes (Park & Lee, 2018).

...

⑦ Gut microbes are important components in the physiology of the gastrointestinal tract (Sommer & Bäckhed, 2013), which play critical roles in the nutritional pretreatment, assimilation and energy harvesting of foods (Ghosh et al., 2014). Understanding the structure of the gut microbiota could help to explore the relationship between intestinal microflora and host health. In order to further evaluate the function of FL, the effect of FL on the gut microbiota was investigated. ⑧ HTS was used to analyze the composition of gut microbiota in mice, and the results of the gut microbiota community revealed that there were significant differences between NC mice and FL-treated mice. Compared with the NC group, the relative abundance of the Firmicutes phylum and the Faecalibaculum, Lactobacillus, and Akkermansia genera in the FL-treated group were all increased, while a decrease in the relative abundance of Bacteroidetes phylum was observed. ⑨ The enhanced relative abundance of Firmicutes was attributable to the improvement of the relative abundance of Faecalibaculum and Lactobacillus. In the gut, short-chain fatty acids (SCFAs) are one of the most important microbial metabolites (Yan et al., 2019), which have been reported to play critical roles in regulating the host's appetite and metabolism (Lin et al., 2012). Several studies reported that Faecalibaculum is positively correlated with SCFAs production (Yan et al., 2019; Zhang et al., 2015). As is well known, Lactobacillus and Akkermansia are important probiotic bacterial genera, which have the ability to generate substances to inhibit pathogens or act against metabolic diseases, thus regulating the balance of gut microbes and improving host health (Cani & de Vos, 2017; Choque Delgado & Tamashiro, 2018; Linares et al., 2017). ⑩ However, since the amount of samples used in this study to analyze the intestinal microorganisms in mice is not large enough, the result could only reflect the composition of gut microbiota in mice to a certain extent.

⑪ In summary, fermentation with *L. casei* was an effective method that dramatically increased the content of total phenolic, total flavone, and exopolysaccharide in litchi juice. The data from animal experiment revealed that intake of FL could improve the immunomodulatory activity of mice by enhancing immune organs indexes (spleen, thymus). Furthermore, FL had the ability to regulate the gut microbiota structure. It significantly raised the relative abundance of beneficial bacteria (Faecalibaculum, Lactobacillus, and Akkermansia). ⑫ Based on the results of this study, we proposed that probiotic litchi juice could be used as a potential function food for improving immunomodulatory function and modulating gut microbiota composition of the host.

a. Restating research questions or methods ____________
b. Summarizing major findings ____________
c. Offering explanations ____________
d. Comparing with previous studies ____________
e. Discussing implications ____________
f. Indicating strengths ____________
g. Indicating limitations ____________
h. Suggesting further research ____________

Discussion 3

Title of the paper: *Tobacco Rotated with Rapeseed for Soil-borne Phytophthora Pathogen Biocontrol Mediated by Rapeseed Root Exudates*

① Tobacco monoculture leads to outbreaks of black shank in tobacco and decreased yields (Kong et al., 1995; Gallup et al., 2006). Our 4-year field experiment confirmed that rapeseed and tobacco rotation significantly suppressed black shank disease in tobacco. ② Many other studies also support the high potential for using rapeseed in rotation, cover, or green manure crops for the suppression of soil-borne diseases (Larkin and Griffin, 2007; Larkin et al., 2010). ③ Reportedly, crop rotation can help reduce soil-borne pathogens, such as fungi, bacteria, oomycetes, and nematodes, by (i) interrupting or breaking the host-pathogen cycle; (ii) altering the soil characteristics to make the soil environment less conducive for pathogen development or survival, often by stimulating microbial activity and diversity or beneficial plant microbes; or (iii) directly inhibiting pathogens either through the production of inhibitory or toxic compounds in the roots or plant residues or the stimulation of specific microbial antagonists that directly suppress the pathogen inoculum (Larkin et al., 2010; Larkin, 2015).

④ In the present study, rapeseed root exudates play an important role in the suppression of *P. parasitica var. nicotianae*. This pathogen is a typical soil-borne pathogen that infects plants through the production of zoospores, which involves a pre-penetration process of zoospore taxis, encystment, cystospore germination, and orientation of the germ tube (Erwin and Ribeiro, 1996). We found that rapeseed roots could attract zoospores to their surface where they were encysted into cystospores. After the cystospores germinated, the growth of the germ tube also proceeded toward the roots. ⑤ Some studies have indicated that the attraction of plant roots to zoospores is not host specific (Deacon, 1988). The chemotaxis and electrotaxis of zoospores toward plant roots are involved in the attraction of zoospores to host and non-host roots (Cameron and Carlile, 1978; Carlile, 1983; van West et al., 2002). After being attracted by the rapeseed roots, some of the spores could not germinate, and some spores even ruptured, which indicates that rapeseed roots can attract zoospores and hyphal growth and then secrete antimicrobial substances against the infection by the zoospores and hyphal growth.

⑥ Evidently, rapeseed root exudates showed dose-dependent antimicrobial activity against the mycelial growth of *P. parasitica var. nicotianae*. A library of antimicrobial compounds, including 2-butenoic acid, valeric acid, 4-methoxyindole, cyclohexyl isocyanate, benzothiazole, 2-(methylthio)-benzothiazole and 1-(4-ethylphenyl)-ethanone, were further identified by GC-MS in the rapeseed root exudates. These compounds showed significant dose-dependent antimicrobial activity against zoospore motility, cystospore germination, and mycelia growth. Notably, 2-butenoic acid, benzothiazole, 2-(methylthio)-benzothiazole, 1-(4-ethylphenyl)-ethanone, and 4-methoxyindole showed antimicrobial activity against zoospore motility and cystospore germination at the concentrations detected in the rapeseed root exudates. ⑦ Among these compounds, benzothiazole and 2-(methylthio)-benzothiazole have previously been identified as antimicrobial compounds against some fungal and oomycete phytopathogens in the root exudates and volatiles of plants as well as in soil bacteria (Bjostard and Hibbard, 1992; Sun et al., 2003; Lane et al., 2004; He et al., 2005; Gaquerel et al., 2009; Hu et al., 2009; Yadav et al., 2011; Zhang et al., 2011; Xu et al., 2012; Yang et al., 2014). 2-Butenoic acid and valeric acid are important organic acids in plant root exudates and possess antimicrobial activity (Corsetti et al., 1998; Baudoin et al., 2003; Sandnes et al., 2005). 1-(4-ethylphenyl)-ethanone exhibited antimycobacterial activity (Rajabi et al., 2005), and 4-methoxyindole and other indole derivatives have been reported in root exudates with antimicrobial activity (Seal et al., 2004; Paudel et al., 2012; Wu and Chen, 2015). The Brassicaceae family produces sulfur compounds that break down to produce isothiocyanates that are toxic to many soil organisms (Sarwar et al., 1998). ⑧ In this study, cyclohexyl isocyanate was identified in the root exudates and showed antimicrobial activity. ⑨ Although the antimicrobial activity of these compounds in rapeseed root exudates was identified, the secretion mechanism of these compounds from root tissue and their mode of action against *P. parasitica var. nicotianae* are still unknown.

⑩ The above data demonstrated that rapeseed can secrete many antimicrobial compounds through its root exudates to kill soil-borne pathogens. Plants are a primary driver of changes in soil microbial communities, and recent studies have documented that crop rotations can dramatically affect these communities (Lupawayi et al., 1998; O'Donnell et al., 2001; Larkin, 2003, 2008; Sturz and Christie, 2003; Larkin and Griffin, 2007; Van Elsas and Costa, 2007; Larkin et al., 2010; Chaparro et al., 2012) and stimulate specific microbial antagonists that directly suppress pathogen inocula (Mazzola et al., 2001; Garbeva et al., 2004; Welbaum et al., 2004). Brassica crops rotated with potato had higher microbial activity and diversity, which may help to suppress soil-borne diseases of potatoes (Garbeva et al., 2004; Welbaum et al., 2004; Ghorbani et al., 2008; Larkin, 2008). ⑪ Thus, whether rotating rapeseed with tobacco increases the soil microbial activity and diversity, the populations of plant-beneficial organisms, and the antagonism toward pathogens to result in disease suppression is interesting and should be evaluated further.

a. Restating research questions ____________

b. Summarizing major findings ____________

c. Offering explanations ____________

d. Comparing with previous studies ____________

e. Discussing implications ____________

f. Indicating strengths ____________

g. Indicating limitations ______________

h. Suggesting further research ______________

2. Sentence structures frequently used in a Discussion

In each move of a Discussion, there are some sentence structures frequently used, which are helpful in writing a Discussion or identifying the elements of a Discussion.

Table 6.2 Sentence Structures Frequently Used in a Discussion

Moves	Sentence structures
Move 1 (Restating research questions or hypothesis)	The objective of the research was to find out/identify ... This study examined if ... This study sets out with the aim of ... It was hypothesized that ... However, ... The hypothesis that ... was generally confirmed. As mentioned in the Introduction section, the present study was designed to ...
Move 2 (Summarizing major findings)	The ... analyses revealed that ... The current/present study showed/indicated/confirmed that ... The results observed support the idea that ... We found no association with respect to ... It has been demonstrated that ... One unexpected finding was that ...
Move 3 (Offering explanations)	There may be several/possible explanations for this. The observed increase/decrease in ... was likely due to ... It seems possible that these results are caused by ... It appears that ... must have led/contributed to the ... The expected results can be attributed to ...
Move 4 (Comparing with previous studies)	This was consistent with the findings of the previous studies in which ... XXX found similar results to our findings in their sample of ... This was in agreement/line with the findings of XXX et al. (*year*) who demonstrated that ... XXX et al. (*year*) previously reported similar results in ... ... have been reported in previous researches, so the trend observed in this study is not unusual. These results/data are inconsistent with prior researches indicating that ... The findings of the current study contradict / don't support the explanation offered in previous studies.

(To be continued)

(Continued)

Moves	Sentence structures
Move 5 (Discussing implications or contributions)	The present results are significant in two aspects This finding has two implications for ... The finding suggests that ... It can be suggested that ... The contribution of this paper is ... It could theoretically enhance the association between ... and ... It is possible/likely/probable, therefore, that ...
Move 6 (Indicating strengths and/or limitations)	There are several strengths to the current study. The present study makes important contributions to the existing knowledge. This study offers some insights into ... This is the first study to report ... This study does have some limitations/weaknesses. The limitations of this study are clear ... However, with a small size, caution must be taken, as the findings might not be transferrable/applicable to ... The findings of the present study are restricted to ...
Move 7 (Suggesting further research)	Studies would be needed to further investigate if ... Further studies on ... are therefore recommended. Much work needs to be done to investigate/explore ... Additional studies are needed to understand moderating factors that influence ... Further research is needed to develop effective strategies for ...

Task 3 **Reread the Examples 1&2 at the begining of 6A and find out the expressions which can help you to identify the moves.**

Moves	Example 1	Example 2
Restating research questions or hypothesis		
Summarizing major findings		
Offering explanations		
Comparing with previous studies		
Discussing implications		
Indicating strengths and/or limitations		
Suggesting further research		

6B Organization: *Specific-general (SG)*

1. General-specific (GS) vs. Specific-general (SG)

In English academic papers, most paragraphs or texts are developed deductively, from more general to specific. This pattern is comparatively simple to read as well as to write, as starting with a generalization or purpose statement constructs a context that provides clear clues for composing or understanding the rest of the paragraph.

In some cases, the paragraph may be arranged in a specific-general pattern, presenting the information in an inductive way. This pattern gradually generalizes what has been found and involves the reader in a process of discovery, arousing their interest and curiosity. While most texts are developed in a simpler GS mode, presenting the topic sentence at the very beginning, SG paragraphs serve as a variation that requires the readers to read more actively.

Task 1 **Read the following paragraphs from Discussions in 6A and finish the exercises.**

Text 1 (Para. 6, *Example 1*)

① This study does have some limitations. ② The four cities included in the sample are urban, low-income, and with high minority populations. ③ The results therefore, are likely to be generalizable to similar communities. ④ Because data on advice on losing weight and eating behaviors were **self-reported** and only collected once, there is a possibility of same-source bias and **underreporting** of unhealthy eating behaviors (Goris et al., 2000). ⑤ The specific weight loss advice **given by** the HCP is also unknown. ⑥ The observed significant differences in consumption behaviors between those **obtaining HCP's advice** and those not getting similar advice were modest. ⑦ In addition, while the classification process for food outlets was robust, it was based on purchased data and there is a possibility that some food outlets were missing from the databases or the information was **outdated** (Powell et al., 2011). ⑧ Analyses of the food environment only accounted for one type of food outlet at a time. ⑨ Lastly, the data were **cross-sectional** and, therefore, the **depicted** associations should not be construed as **cause-effect relationships**.

1) Is GS or SG pattern used in this paragraph?
2) Which sentence is the topic sentence?
3) How many limitations are mentioned? In which sentences are they located respectively?

Text 2 (Para. 2, *Discussion 3*)

① In the present study, rapeseed root exudates play an important role in the suppression of *P. parasitica var. nicotianae*. ② This pathogen is a typical soil-borne pathogen that infects plants through the production of zoospores, which involves a pre-penetration process of zoospore taxis, encystment, cystospore germination, and orientation of the germ tube (Erwin and Ribeiro, 1996). ③ We found that rapeseed roots could attract zoospores to their surface where they were encysted into cystospores. ④ After the cystospores germinated, the growth of the germ tube also proceeded toward the roots. ⑤ Some studies have indicated that the attraction of plant roots to zoospores is not host specific (Deacon, 1988). ⑥ The chemotaxis and electrotaxis of zoospores toward plant roots are involved in the attraction of zoospores to host and non-host roots (Cameron and Carlile, 1978; Carlile, 1983; van West et al., 2002). ⑦ After being attracted by the rapeseed roots, some of the spores could not germinate, and some spores even ruptured, which indicates that rapeseed roots can attract zoospores and hyphal growth and then secrete antimicrobial substances against the infection by the zoospores and hyphal growth.

1) Is GS or SG pattern used in this paragraph?
2) Choose the suitable information marked a–f to complete the form that follows.
   a. Secretion of antimicrobial substances against infection
   b. Rapeseed root exudates help to suppress the pathogen
   c. Zoospores attraction and encystment
   d. Comparison with other studies of the taxis of zoospores
   e. Infection process of the pathogen
   f. Cystospore germination and orientation of the germ tube

Major finding (Topic): (1) ____________________
Explanation

(To be continued)

(Continued)

Principle: (2) ________________
Step 1: (3) ________________
Step 2: (4) ________________
Supporting evidence: (5) ________________
Step 3: (6) ________________

Text 3 (Para. 3, *Example 2*)

① The effects of the concentration of Au on soil respiration were variable in Au-amended soils. ② In some soils, such as MNP and to a certain extent PPN soils, increasing metal concentration did not always lead to a substantial reduction in soil respiration rates, suggesting that a lower contamination threshold was required for detectable impairment of soil functionality. ③ The effects of Pt amendment in the same soil types were different, with stimulation or inhibition of soil respiration largely correlated with increasing metal concentration. ④ This demonstrated that soil microbial communities responded differently to Au and Pt contamination, which could reflect the different mobility, adsorption and bioavailability of these heavy metals in different soil types. ⑤ In summary, the hypothesis that the presence of Au and Pt would affect soil respiration differently was generally confirmed. ⑥ However, the response was varied with both soil type and metal concentration, highlighting the complexity of the interaction of these heavy metals with soils (Chander and Brookes, 1991; Killham, 1985).

1) Is GS or SG pattern used in this paragraph? ________________

2) Analyse the organization of the paragraph and complete the following form.

Sentences	Organization (general statement / specific detail)	Functions
①		Finding1:
②		
③		Finding 2:
④		Interpretation to the above findings
⑤		
⑥		Further explanation

2. SG model above the paragraph level

As stated in Unit 1, the GS model can be used at the paragraph level as well as for larger units of discourse, such as a series of paragraphs in a section or even the whole text, and so is the SG pattern.

The Introduction section, for example, is often arranged in the GS model, gradually moving from general ideas toward one or more specific research focuses. The Discussion section, however, usually "zooms out" from the research details of the Results and takes a broad look at the major findings and the whole study. Therefore, researchers tend to use the SG model to organize their ideas in Discussion.

Task 2 **Reread Example 1 and complete the following diagram according to the information given. This will help you understand how this Discussion moves from specific to general ideas.**

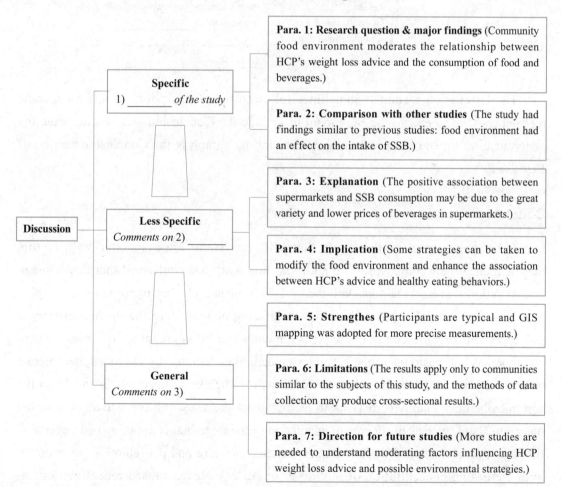

Discussion

- **Specific**
  1) ________ *of the study*
  - **Para. 1: Research question & major findings** (Community food environment moderates the relationship between HCP's weight loss advice and the consumption of food and beverages.)
  - **Para. 2: Comparison with other studies** (The study had findings similar to previous studies: food environment had an effect on the intake of SSB.)
- **Less Specific**
  Comments on 2) ________
  - **Para. 3: Explanation** (The positive association between supermarkets and SSB consumption may be due to the great variety and lower prices of beverages in supermarkets.)
  - **Para. 4: Implication** (Some strategies can be taken to modify the food environment and enhance the association between HCP's advice and healthy eating behaviors.)
- **General**
  Comments on 3) ________
  - **Para. 5: Strengthes** (Participants are typical and GIS mapping was adopted for more precise measurements.)
  - **Para. 6: Limitations** (The results apply only to communities similar to the subjects of this study, and the methods of data collection may produce cross-sectional results.)
  - **Para. 7: Direction for future studies** (More studies are needed to understand moderating factors influencing HCP weight loss advice and possible environmental strategies.)

Task 3 Reread Discussion 2 and complete the following diagram.

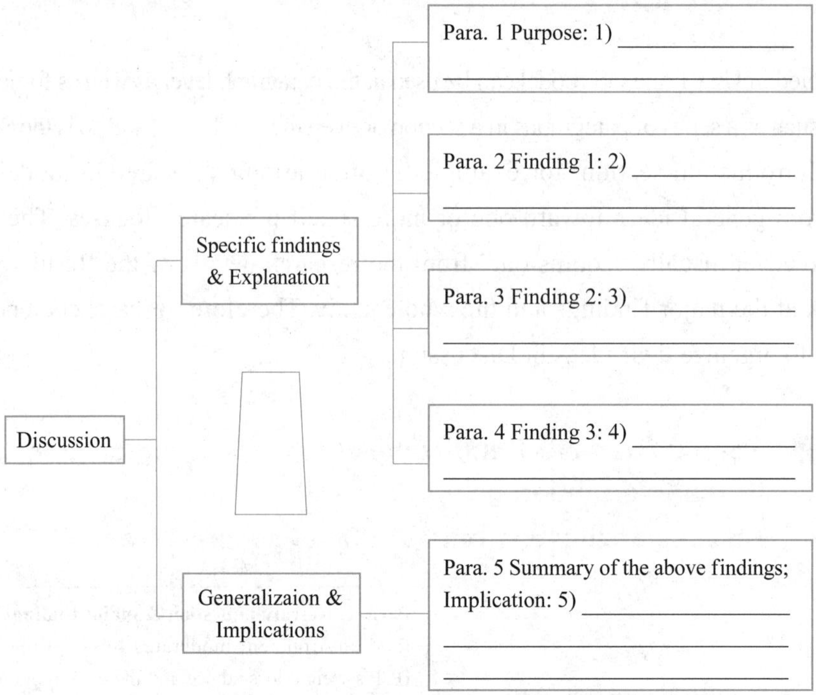

The Discussion section is sometimes followed by a Conclusion. In this case, some researchers tend to extend the SG model to include the Conclusion section and leave the generalization for the last section. The following paragraph is the Conclusion section of *Example 2*.

Conclusion

It has been shown in this study that soil ecosystems are associated with factors that influence their composition. In addition, this work has confirmed that heavy metal concentration represents another key factor. Heavy metals play an important but complex role in soil microbial communities by accelerating or inhibiting the decomposition of polymers and organic compounds. Heavy metals can be aggregated into minerals, or incorporated into basic microbial secondary metabolites secreted by microbes that interact with metal surfaces (Bahadur et al., 2016). However, the exact impact of Au and Pt on the soil microbial community is difficult to predict given the array of biotic and abiotic factors unique to a soil, such as the degree of adaptation of the microbial community and vegetation (biotic), together with soil organic matter content, moisture and pH (abiotic). Amendment with increasing concentration of mobile Pt or Au complexes caused reductions in soil respiration rates in most tested soils, demonstrating that both Au and Pt complexes are toxic to soil microbial communities. However, chemical analyses and specification studies would be needed to further investigate if ions or complexes cause toxicity. These reactions were related to the soil type and the concentration of the amendment. While mobile Au appears to be somewhat more toxic to soil microbial communities, toxicity levels appear to be linked to soil properties.

In the Discussion section, the writer states the major finding of the study and the specific factors affecting the impact of heavy metals on soil respiration. The Conclusion section summarizes the findings and factors and depicts a broader view of the study. The diagram below graphically illustrates the design of these two sections.

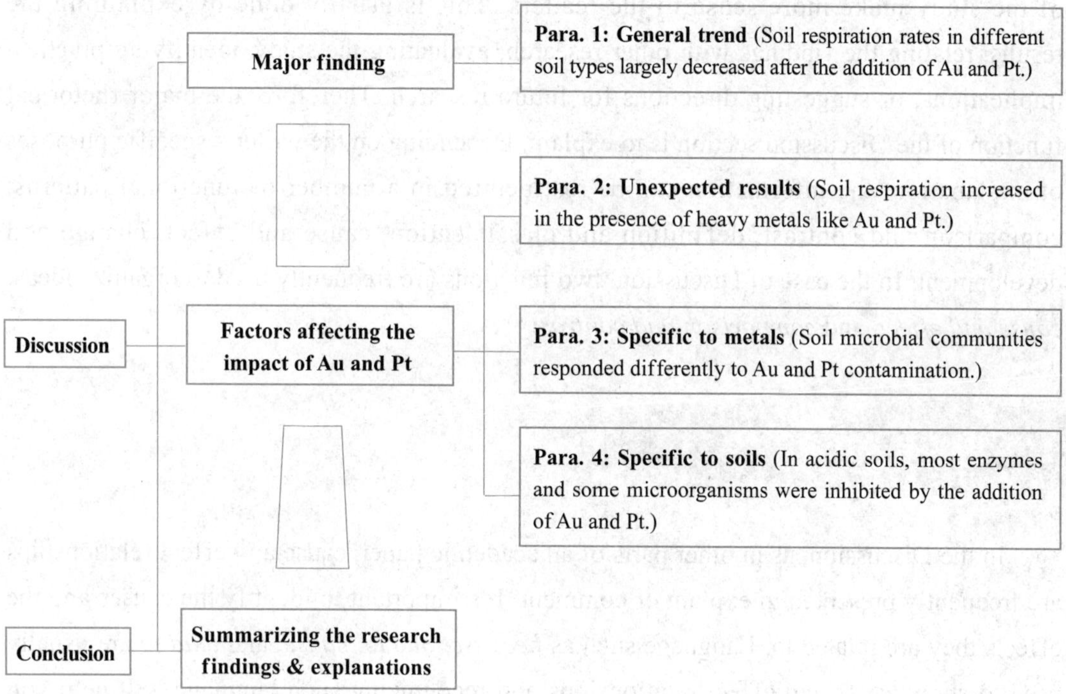

6C Rhetorical Function: *Cause and Effect; Comparison and Contrast*

While reading a text, identifying its rhetorical functions is useful to recognize relationships between ideas and how they change throughout the text. In this unit, we'll analyze the functions in a Discussion.

The Discussion section is intended to interpret the major findings so that the results of the study make more sense to the readers. This is usually done by explaining the results, relating the findings with other research, evaluating the study, identifying practical implications, or suggesting directions for future research. Therefore, the major rhetorical function of the Discussion section is to explain. Depending on the writer's specific purposes of explanation, the information can be restructured in a number of functional patterns: comparison and contrast, definition and classification, cause and effect, change and development. In the case of Discussion, two functions are frequently used to organize ideas: *cause and effect*, and *comparison and contrast*.

1. Cause and effect

In the Discussion, as in other parts of an academic paper, cause and effect relationships are frequently presented to explain or comment. It's important to identify the causes and the effects they are related to. Language such as *because, due to, cause*, and *lead to* are usually used to show cause and effect relationships and recognizing such language will help you understand these relationships.

Example 1

Veterinary drug treatment is often challenging **due to** size and body mass variations, as well as pharmacokinetic inter-species variability. The availability of approved veterinary products is relatively small, **consequently leading to** the need for extemporaneous manufacturing.

In this example, the phrase *due to* indicates the cause of the challenges in veterinary drug use. The adverb *consequently* and the phrase *lead to* are used to express the effect of the lack of approved veterinary products.

Task 1 **Read the following sentences, underline the words or phrases that show cause and effect relationships and identify how causes and effects are arranged by choosing from the modes given in the box.**

Cause and effect relationships can be expressed in different ways:

a. Cause → Effect

b. Effect → Cause

c. Cause → Effect (Cause) → Effect

1) The residue at SEREC may have mineralized faster because of the warmer climate experienced between cropping systems and the higher soil microbial activity of the clay soil.________
2) This large gap in previous soybean productivity and harvest dates resulted in differing amounts of residue and time of decomposition. ________
3) The positive association between supermarkets and SSB consumption may be due to the variety of beverages found hin supermarkets. ________
4) A sudden lack of mobility across borders and within countries has caused labour shortages in countries that are reliant on seasonal migrant workers in the agri-food sector, which, in turn, has affected food availability and prices globally (FAO 2020). ________
5) Although the difference in temperature was not large, it may have led to more mineralization throughout the season, resulting in the accumulated differences seen at SEREC and not at PTRS. ________
6) The increased rice grain yield recorded when following a 5.4 MG soybean can be explained by the elevated soybean plant population instead of the crop management. ________
7) Food consumption is generally quite inelastic and it takes several years for production to adjust fully to a price change, so the GDP shocks only have a modest impact on global

production and consumption. ________

8) Altering the soil characteristics could make the soil environment less conducive for pathogen development or survival, often stimulating microbial activity and diversity or beneficial plant microbes. ________

Task 2 **Complete the following sentences with the appropriate words or phrases in the list. Change their forms or add functional words if necessary.**

because of	cause	due to	lead (to)	result (in/from)

1) According to our results, the sharp decline in economic growth ________ a decrease in international meat prices by 7%–18% in 2020 and dairy products by 4%–7% compared to a business as usual situation.
2) Although the income losses and local supply chain disruptions associated with the pandemic undoubtedly ________ an increase in food insecurity in many developing countries, global food consumption is largely unaffected ________ the inelastic demand of most agricultural commodities and the short duration of the shock.
3) The orodispersible film needs to stick to the mucosa and not float around in the mouth. A non-adhesive film is easier for the animal to spit out, ________ treatment failure.
4) The altered Zmorph printer had difficulties extruding the ink in a constant manner ________ the circling motion of the stepper motor, ________ high variations in weight, thickness, and drug amounts.

Task 3 **Reread at least two discussions in 6A and find some examples of the two functions discussed in this section.**

Model: Organic matter can also adsorb some heavy metals, forming stable complexes and rendering such metals non-bioavailable to the soil microbiota (Hamid et al., 2020). Therefore, the organic matter in FLN soil was probably beneficial to soil respiration in that it adsorbed most of the introduced Au while making available ample resources for microbial activities (respiration) (Lai et al., 2013). Increased soil respiration in the presence of heavy metals such as Pt and Au could also occur as a result of a stress response, caused by increased diversion of energy from growth to maintenance energy requirements, different pH, and heavy metal bioavailability. (*Para. 2, Example 2 in 6A*)

2. Comparison and contrast

When writers want to explain a new idea in terms of something familiar, or to evaluate items and choose the best one, they compare the items to show how they are both similar and different. Writers contrast by showing only how items are different. Their purposes can be to define a concept by saying what it is not, to correct mistaken assumptions, or to show the opposite side of an argument (Argent & Alexander, 2013). They help to explain connections, significance, advantages or disadvantages, distinctive features, etc. Some vocabulary and structures are used to compare and contrast.

Example 2

Prednisolone is a glucocorticoid used in the treatment of arthritis, asthma, skin disorders and allergic dermatoses in **both** cats and dogs. Dogs are dosed with 0.5–1 mg/kg/day prednisolone. **However**, there is evidence that cats have **fewer** glucocorticoid receptors and express a **higher** resistance to glucocorticoid treatment; therefore, cats require **larger** doses **compared to** dogs and are treated with doses of 1–2 mg/kg/day. At present, prednisolone tablets can only be found in the strengths of 5, 20, and 40 mg in Finland.

Example 3

European Pharmacopoeia defines extemporaneous preparations as pharmaceutical preparations that are individually prepared for a specific patient or patient group and supplied to the patient after preparation. **In contrast**, stock preparations are prepared in bulk in advance and stored until needed.

In the two examples, with the bold-typed expressions, the author compares two things (cats vs. dogs; extemporaneous preparations vs. stock preparations) in order to find out their similarities or differences.

Task 4 **Read the following sentences and decide whether the italicized words or phrases show similarities (comparison) or differences (contrast). Write S for similarity and D for difference.**

1) Significant differences in rice grain yield appeared on ***both*** silt loam and clay soils in 2018. ________
2) Residue left on the soil surface will decompose ***slower than*** incorporated residue. ________
3) All response variables were analyzed as the previous MG strip acting as one plot, ***similar to*** the statistical analysis of Ortel et al. (2020). ________
4) The association between HCP's advice on losing weight and consumption of sugar-sweetened beverages and fast food will be ***weaker*** among those who live closer to convenience stores and limited-service restaurants, ***compared with*** those who do not live near these food outlets. ________
5) Unhealthy food outlets were closer to participants' homes, ***while*** healthy food outlets were farther from participants' homes. ________
6) A ¼ mile distance was used for LSR and convenience stores, as there was limited variability in the distribution of these food outlets beyond a ¼ mile. ***Conversely***, a ½ mile distance was used for small grocery stores and supermarkets, as there was limited variability in the distribution of these outlets within ¼ mile. ________
7) Reduction in soil respiration and undesirable changes in soil matrix biochemistry have been correlated with Cd, Zn and Cu soil pollution. ***Similarly***, amendment of soil microcosms from several sites in Australia led to significant changes in the bacterial community. ________
8) The rice crop aboveground TNU was significantly influenced by the previous soybean management in 2018 on the silt loam soil but ***not*** on the clay soil. ________
9) The trend seen in 2018 rice grain yield of the previous soybean MG ***follows the same general trend as*** the 2017 soybean crop plant population. ________
10) In the case of the sample measurements, 200 μL of the substrate solution (0.2 mM) was added ***instead***. ________

Task 5 **Read the following sentences, underline the words or phrases that show similarities or differences and write them in the form that follows.**

1) The large gap in previous soybean productivity that occurred at SEREC also occurred at PTRS.
2) At a suboptimal N rate, the rice crop following optimum-planted soybean measured 61% apparent N fertilizer recovery (ANFR), whereas the rice crop following late-planted soybean measured 80.5% ANFR.
3) The increased rice grain yield recorded when following a 5.4 MG soybean can be explained by the elevated soybean plant population instead of the crop management.
4) Laska et al. found similar results to our findings in their sample of adolescents: those having a fast food restaurant or convenience store within 1,600 m of home consumed 25% and 24% more SSB, respectively.
5) Our result of lower reported SSB consumption in the absence of a supermarket also align with the findings from Gustafson et al. who found that those who shopped frequently at a supermarket had a higher odds of consuming SSB.
6) GIS mapping was used to determine outlet proximity to each participant's home rather than grouping participants by census tract, allowing for more precise measurements of the presence and absence of food outlets.
7) The initial soil respiration rates differed substantially among the five soils, with BGR soil having the highest CO_2 production, followed by JBR, FLN, MNP and PPN.
8) It appears that pathogens are unable to effectively degrade mature suberin (Kolattukudy, 1981, 1984), and the same is likely to be true for ligno-suberin complexes.

Functions	Expressions
Comparison	also, ...
Contrast	

Task 6 **Complete the following sentences with the appropriate words or phrases in the list. Capitalize the first letter if necessary.**

compared with	conversely	in contrast to	lower than
rather than	similar	similarly	while

1) ________ these first two reactions of tissues to wounding (i.e. as physical objects or as biochemical solutions), the third response (physiologically active tissues) involves the active participation of the tissue.
2) Maize is a feedstock for ethanol production, so it is especially affected by oil price variability. ________, vegetable oil is the major feedstock for biodiesel, so its price is also sensitive to oil price volatility.
3) Legume crops have relatively high N concentrations in their unharvested biomass ________ other agronomic crops.
4) Rice yield was consistently higher when rice followed a soybean treatment ________ a fallow strip.
5) On the addition of Pt, NAG, PHOS and XYL activities increased at most tested concentrations ________ other enzymes were inhibited.
6) In PPN soil samples, the initial soil respiration rate was ________ the respiration rate observed in the other four soil types.
7) In the BGR soil, Au soil spiking inhibited the activities of all seven enzymes at all the concentrations tested. ________, Pt increased NAG enzymatic activity by up to 40% to 80% at different concentrations.
8) The addition of different concentrations of Au caused a significant decrease in soil respiration in most of the tested concentrations, with a ________ trend being observed in soils amended with Pt.

6D Register: *Hedging*

In academic writing, authors need to be cautious with the language they use when explaining their findings, drawing conclusions, or presenting implications in the Discussion section. They need to avoid expressing absolute certainty, where there may be a small degree of uncertainty, and to avoid making over-generalization, where a small number of exceptions might exist. In this case, hedging, the use of cautious language, helps to reduce the force of a statement or interpretation, or to distance the authors from the claims or findings, thus making them more acceptable.

At the same time, it would be deemed as being abrupt and unreliable if an argument or conclusion is made without providing empirical or literature evidence. Hedging expressions like *perhaps*, *may*, *seem* and *likely*, could make the claim or explanation more flexible, showing that the author is speculating about something and is not completely certain. They tend to transform a fact into a guess, signalling the authors' interpretation or suggestion.

Now read the following text (paragraphs 3 & 4 in ***Example 2*** in 6A) and pay attention to the hedging devices used and the effects they have.

Example

The effects of the concentration of Au on soil respiration were variable in Au-amended soils. In some soils, such as MNP and to a certain extent PPN soils, increasing metal concentration did not always lead to a substantial reduction in soil respiration rates, suggesting that a lower contamination threshold was required for detectable impairment of soil functionality. The effects of Pt amendment in the same soil types were different, with stimulation or inhibition of soil respiration largely correlated with increasing metal concentration. This demonstrated that soil microbial communities responded differently to Au and Pt contamination, which could reflect the different mobility, adsorption and bioavailability of these heavy metals in different soil types. In summary, the hypothesis that the presence of Au and Pt would affect soil respiration differently was generally confirmed. However, the response was varied with both soil type and metal concentration, highlighting the complexity of the interaction of these heavy metals with soils (Chander and Brookes, 1991; Killham, 1985).

In acidic BGR soils, all the tested enzymes were inhibited by the addition of Au and Pt (except NAG at 1, 25, 100 mg/kg Pt concentration), suggesting that, under acidic conditions, Au and Pt were toxic to most enzymes. Previous studies have shown similar findings under acidic conditions, with Cu inhibiting enzyme function (Gupta and Aten, 1993; Romié et al., 2014; Tyler and Mcbride, 1982). Acidic environments can also lead to decreased growth in some microorganisms as more metabolic energy is used for maintenance instead of respiration and enzyme function (Sherameti and Varma, 2010). This may also be related to the mobility of Pt and Au under acidic conditions.

In the above example, the text mainly consists of the author's explanation of the results and findings. The underlined words are hedges, which can soften the certainty of claims, explanations or arguments. By hedging, the researchers are cautious about what they say and leave room for alternatives. In this way, their claims and arguments sound more reliable and acceptable. More hedging expressions are listed in the following table.

Table 6.3 Hedging Devices

Patterns	Expressions
Verbs and phrases of uncertainty	seem, tend to, look like, appear, think, feel, believe, view
Tentative verbs	suggest, estimate, assume
Modal auxiliary verbs	may, can, would, could, might
Adverbs of frequency and degree	usually, sometimes, often, occasionally, frequently, generally, infrequently, rarely, seldom, about, approximately, roughly, somehow, somewhat
Modal adverbs and adjectives	maybe, perhaps, possibly, probably, virtually, apparently, conceivably, presumably, arguably; possible, probable, likely, unlikely

(To be continued)

(Continued)

Patterns	Expressions
Modal nouns	possibility, probability, likelihood, assumption, claim, estimate, suggestion
Expressions of quantity	some, many, few, a few, little, a lot of, a number of, lots of, most, the majority of
Weak verbs	reduce (cf. prevent), influence (cf. cause), show (cf. demonstrate), report (cf. prove)

Task 1 Compare the two versions of each statement and discuss with your partner about the differences the hedges make.

1a. The development has benefited from the recent technology upgrade.

1b. The development seems/appears to have benefited from the recent technology upgrade.

2a. The positive association between supermarkets and SSB consumption is due to the variety of beverages found in supermarkets.

2b. The positive association between supermarkets and SSB consumption may be due to the variety of beverages found in supermarkets.

3a. A study in São Paulo, Brazil, found that adults who lived in a census tract with a greater variety of SSB consume more SSB.

3b. A study in São Paulo, Brazil, found that adults who lived in a census tract with a greater variety of SSB were more likely to consume SSB.

4a. However, there was no study on the health function of probiotic litchi juice.

4b. However, there were few studies on the health function of probiotic litchi juice.

Task 2 Read the following passage and fill in the blanks with the expressions given in the box. Capitalize the initial letter if necessary.

appear	approximately	may	might	on average
possible	possibly	some	tend to	would

Title of the paper: *Sweetpotato Cultivars Differ in Efficiency of Wound Healing*

According to the regression models, the probability that wounds heal decreases with increasing dry matter (DM) content, and, 1) ________, if the dry matter content increases by

1%, the lignification score (L.S.) decreases by 2) ________ 0.05.

It was also observed that the regression model in trial 5 was less steep than in trials 4, 6 and 7. This coincided with a higher relative humidity in trial 5, which was 75.9% as opposed to 71.1%, 67.3% and 64.7% in the other trials. Although the difference in relative humidity did not affect the cultivars with a low dry matter content, it did affect the cultivars with a higher dry matter content. 3) ________ there is an interaction between the DM content of the cultivar and the relative humidity necessary for wound healing.

The mechanism by which the DM content might affect wound healing is not understood, and requires further investigation. It is 4) ________ that wound healing ability is directly related to the rate of desiccation of the tissue. Parameter readings indicate that tissue with a high DM content initially loses water at the same rate as tissue with a low DM content, and 5) ________ therefore reach a critical level of moisture content more rapidly than tissue with a low DM content. The hypothesis 6) ________ be that below a critical level of moisture content the tissue stress results in failure to form the protective lignified layer under the wound.

There would 7) ________ to be several factors affecting the wound healing ability since there were 8) ________ consistent outlying cultivars. The cultivar KSP20 consistently healed less well than predicted from the DM/L.S. relationship, while Caplina and Yarada consistently healed better than expected.

The relationship seen here between L.S. and DM 9) ________ explain previous results obtained where we observed that high dry matter cultivars in Tanzania 10) ________ have a shorter shelf life (Rees et al., 2003).

Task 3 Rewrite the following sentences with the types of hedging devices given in the brackets. You may need to refer to Table 6.3 for help.

1) The expression of these three genes must have led to the generation of DHM and the reduction of cyanidin precursors. (*verbs of uncertainty*)
2) The lack of association between HCP's advice and SSB consumption when participants lived near supermarkets indicates that HCP's advice on losing weight may not be powerful enough to overcome environmental cues such as ease of access, in-store marketing, and price discounts. (*tentative verbs*)
3) However, modifications to the food environment theoretically enhance the association between HCP's advice and healthy eating behaviors. (*modal auxiliary verbs*)

4) Probiotics are added to fruit juices, dairy products or meat products to increase their health benefits. (*adverbs of frequency*)
5) Food consumption is quite inelastic and it takes several years for production to adjust fully to a price change. (*adverbs of degree*)
6) Therefore, the observed decrease in soil respiration rates in this study was due to Au and Pt toxicity to soil microbial groups. (*modal adverbs*)
7) The results therefore, are generalizable only to similar communities. (*modal adjectives*)
8) Because data on advice on losing weight and eating behaviors were self-reported and only collected once, there is same-source bias and underreporting of unhealthy eating behaviors. (*modal nouns*)
9) Other studies also support the high potential for using rapeseed in rotation, cover, or green manure crops for the suppression of soil-borne diseases (Larkin and Griffin, 2007; Larkin et al., 2010). (*expressions of quantity*)
10) Several studies proved that crop rotation can help reduce soil-borne pathogens, such as fungi, bacteria, oomycetes, and nematodes (Yan et al., 2019; Zhang et al., 2015). (*weak verbs*)

6E Cohesion (5): *Lexical Reiteration*

Lexical reiteration is an important aspect in lexical cohesion, and provides a basis for discourse coherence. It creates cohesion in a text by repeating lexical items or key ideas, which are not necessarily the same word or phrase. According to Halliday and Hasan (2001), lexical reiteration falls into four categories: repetition of the same word or phrase, use of a synonym, use of a superordinate, or use of a general noun. These four categories provide options to achieve cohesion that range from the most specific (use of the same word) to the most general (use of a general noun).

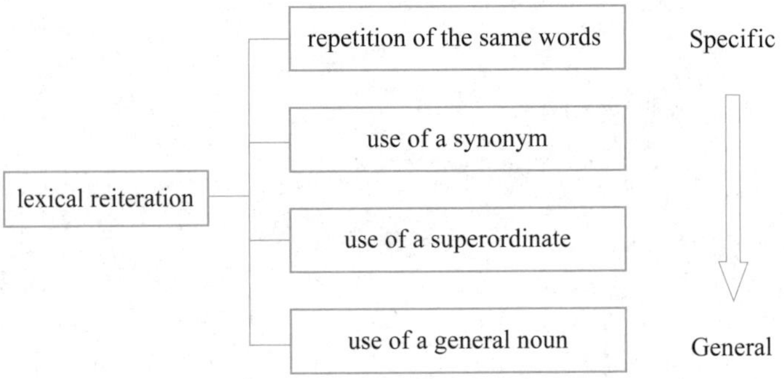

Figure 6.1 Lexical Reiteration

As the use of general nouns has been discussed in Unit 4, we'll focus on the other three ways of lexical reiteration: repetition, synonym and superordinate.

Example

Title of the paper: *Temporary Reduction in Daily Global* CO_2 *Emissions*

Figure 4 illustrates how the [1]**world prices** of [2]**grains** and [3]**biofuels** in the scenario are [4]**affected** by [5]**variability** in the [1]**international price** of oil. [2]**Maize** is a feedstock for [3]**ethanol** production, <u>so</u> <u>it</u> is especially [4]**affected** by oil [1]**price**

[5]**variability**. Similarly, vegetable oil is the major feedstock for [3]**biodiesel,** so its [1]**price** is also [4]**sensitive to** oil [1]**price** [5]**volatility**.

Key Items	Repetition	Synonym	Superordinate — Hyponym
1. price	price, prices	world prices, international price	
2. grain			grain — maize
3. biofuel			biofuel — ethanol, biodiesel
4. affect	affected	sensitive to	
5. variability	variability	volatility	

In this short paragraph, the author reiterates the five key words in different ways, repeating the exact word or using its synonym or superordinate/hyponym. By tracing back to previous items, reiteration connects ideas located in different places, thus achieving cohesion throughout the whole paragraph. In addition, the underlined words in this example also contribute to the cohesion of the text. Can you figure out what cohesive devices are used?

Task 1 **Read the following paragraph and show how the words in bold are reiterated by arranging them in the table below. The first one is already done as an example.**

Title of the paper: *Temporary Reduction in Daily Global CO_2 Emissions*

The **modest** global **production changes resulting from** the **demand shock** imply that the **effect** on global **GHG emissions** is also **modest**, around 1% in 2020–2021. However, for some of the large **producers**, the **emission reductions**, especially from **animal production**, are in the order of 2%–3%. In absolute terms, these **changes** correspond to around 50 Mt of **CO_2 equivalents** in 2020 and 2021. From a **climate policy** perspective, the **modest impacts** on **agricultural GHG emissions** might seem disappointing. However, it is important to bear in mind that the scenario we consider does not include the **effect** of **the European Green Deal** or any **other policies** that were not implemented in 2019. **Such policies** that **affect production** and **consumption** incentives in the long term, have the potential for an **impact** that is much greater than a (hopefully) short-term **disruption**.

Reiterated items	Repetition	Synonym	Superordinate/ Hyponym	Other ties
modest	modest			

(To be continued)

(Continued)

production				producers
change				
effect				
demand				
GHG emissions				
climate policy				other policies, such policies

Task 2 Analyze the following paragraph and identify the reiterations and other cohesive ties by completing the table below.

The effects of the concentration of Au on soil respiration were variable in Au-amended soils. In some soils, such as MNP and to a certain extent PPN soils, increasing metal concentration did not always lead to a substantial reduction in soil respiration rates, suggesting that a lower contamination threshold was required for detectable impairment of soil functionality. The effects of Pt amendment in the same soil types were different, with stimulation or inhibition of soil respiration largely correlated with increasing metal concentration. This demonstrated that soil microbial communities responded differently to Au and Pt contamination, which could reflect the different mobilities, adsorption and bioavailability of these heavy metals in different soil types. In summary, the hypothesis that the presence of Au and Pt would affect soil respiration differently was generally confirmed. However, the response was varied with both soil type and metal concentration, highlighting the complexity of the interaction of these heavy metals with soils (Chander and Brookes, 1991; Killham, 1985).

Cohesive Ties					
Grammatical cohesion			**Lexical cohesion**		
Conjunction	Substitution	Reference	Lexical reiteration: repetition, synonym, superordinate	Lexical chains	General nouns
				X	

Task 3 Rewrite the following paragraph by replacing the underlined parts with a synonym or superordinate.

In acidic BGR soils, all the tested enzymes were <u>decreased</u> by the addition of Au and Pt (except NAG at 1, 25, 100 mg/kg Pt concentration), suggesting that <u>in acidic soils</u> Au and Pt were toxic to most enzymes. Previous studies have shown similar findings <u>in acidic soils</u>, with Cu <u>decreasing</u> enzyme function (Gupta and Aten, 1993; Romié et al., 2014; Tyler and Mcbride, 1982). <u>Acidic soils</u> can also lead to <u>decreased</u> growth in some microorganisms as more metabolic energy is used for maintenance instead of respiration and enzyme function (Sherameti and Varma, 2010). This may also be related to the mobility of Pt and Au <u>in acidic soils</u>.

6F Assignments: *Reading Comprehension and Vocabulary*

Part one Reading comprehension

Task 1 Read the following text and tell whether the statements that follow are true (T) or false (F).

Gan River Basin 赣江流域

Acipenseridae *n.* 鲟科
Cluoeidae *n.* 鲱科
Anguillidae *n.* 鳗鲡科
Cynoglossidae *n.* 舌鳎科
Teraodontidae *n.* 鲀科

Tenualosa reevesii 鲥鱼

Dams continue to have innumerable impacts on freshwater fishes and ecosystems [2,12]. This study found that major dam construction in the Gan River Basin has resulted in dramatic changes to multiple dimensions of fish diversity. Comparisons between the historical (pre-dam) and current (post-dam) periods showed that the number of total and native species have declined. Most notably, 29 native species were lost in the present, including the extirpation of all species from the families Acipenseridae, Cluoeidae, Anguillidae, Cynoglossidae and Teraodontidae (Tables S2 and S3). By contrast, six alien species were gained.

Many freshwater ecosystems in China have experienced massive fish declines in recent decades; a pattern also observed in this study. Other threats, such as water pollution, overfishing, invasive species, and climate change, have also affected fish diversity in the Gan River Basin. For example, the yield of Tenualosa reevesii rapidly declined from 309–584 t in 1960, 74–157 t in 1970, and 12 t in 1986 due to overfishing [17,66,67]. The continuous input of industrial wastewater and domestic sewage has caused the gradual deterioration of the water quality and indirectly affected fish diversity [67–69].

In many instances, freshwater fish may be under multiple pressures and, at the same time, may act **synergistically**, exposing the species to greater risks [70]. Therefore, further research on the interaction between unknown or poorly understood stressors (pollution, invasive species, and climate change) and multiple stressors deserves further attention.

synergistically *adv.* 协同地

______ 1) This Discussion contains all the moves.

______ 2) The research questions have been restated in this Discussion.

______ 3) This Discussion is organized in *General-Specific* pattern.

______ 4) The findings of this research show that dam construction has negative effect on fish diversity and ecosystems.

______ 5) The continuous input of industrial wastewater and domestic sewage directly affected fish diversity.

______ 6) Further research could be conducted on the relationship between different stressors which we may not know.

Task 2 Read the text and answer the questions that follow.

① Plant **growth regulators** (PGRs) play an important role in flower production, which in small amounts promotes or inhibits or quantitatively **modifies** growth and development (Kumar et al., 2014). ② In the current study, the impact of different PGRs on floral size, weight, and some **antioxidant attributes** of two **cultivars** of African **marigold**, i.e. "Basanthi" and "Narangi", was evaluated. ③ In terms of floral diameter, both cultivars differentially responded to different PGRs. ④ "Basanthi" exhibited a 60% increase in floral diameter under the influence of 150 mg/L **IBA**, while "Narangi" showed its maximum response (48% enhancement) under 300 mg/L **OA**. ⑤ **ABA**, **NAD**, **GA_3**, and **SA** also increased the floral diameter of marigold up to 35%, 48%, 34% and 34%, respectively (Figure 2). ⑥ Riaz et al. (2013), Mitchell and Stewart (1939), and Dhuma et al. (2018) reported similar results in marigold and **tuberose** under the influence of ABA and NAD. ⑦ The

growth regulator 生长调节剂

modify *v.* 调整

antioxidant *adj.* 抗氧化的

attribute *n.* 特性，属性

cultivar *n.* 品种

marigold *n.* 万寿菊

IBA 吲哚3丁酸（indole-3-butyric acid）

OA 草酸（oxalic acid）

ABA 脱落酸（abscisic acid）

NAD n–乙酰噻唑烷（N-acetyl thiazolidine）

GA_3 赤霉酸（gibberellic acid）

SA 水杨酸（salicylic acid）

tuberose *n.* 晚香玉

induce *v.* 引发，诱发
increment *n.* 增量
chrysanthemum *n.* 菊花
assimilation *n.* 吸收
photosynthetic *adj.* 光合作用的
uptake *n.* 吸收，吸收速度

GA_3-**induced increment** in floral diameter was reported earlier in marigold and **chrysanthemum** (Shinde et al. 2010). ⑧ The increase in floral diameter under the influence of SA was stated as the result of increased CO_2 **assimilation**, **photosynthetic** rate and mineral **uptake**, as supported by previous studies (Karlidag et al., 2009). ⑨ IBA plays a vital role in increasing cell division (Shahzad et al., 2002), and hence was found to be a promising way to improve floral diameter of marigold. ⑩ Our results about the influence of IBA and OA on floral diameter of marigold are supported by previous findings in marigold, **red firespike** (Odontonema strictum), and **henna** (LawsoniaInermis) (Kaushik & Shukla, 2020).

red firespike［花卉］鸡冠爵床
henna *n.* 指甲花

Questions

1) How many moves can you find in this Discussion? Identify the sentences that correspond to each move in the table below and determine what the tenses are used in each move.

Moves	Typical elements	Sentences	Tenses
Move 1	Research Questions or Hypothesis		
Move 2	Major Findings		
Move 3	Explanations		
Move 4	Previous Studies		
Move 5	Implications		
Move 6	Strengths and/or Limitations		
Move 7	Further Research		

2) Quite a few previous studies are mentioned in this Discussion. Why did the author do it? Is it common in the RAs in your field?
3) What tense is mainly used in this Discussion? Why?
4) What is the research question of the study?
5) What effect do PGRs have on flower growth of African Marigold?

6) What are the causes for the increase in floral diameter under the influence of SA according to previous studies?

7) This passage is the first paragraph of a Discussion. What would probably be discussed in the following parts of this Discussion?

8) Find out the signal words for comparison & contrast and cause & effect.

Functions (Explanation)	Sentences	Signal words
Comparison & Contrast		
Cause & Effect		

9) Read Sentence ④ and write a sentence of comparison and contrast with the structure "A ..., while B ... ".

10) Explain the causes of global warming using the cause-effect expressions you've found.

Task 3 Write a Discussion section for a research you have been involved in.

Part two Vocabulary

Adjectives are widely used in academic texts when describing, evaluating and classifying concepts. Evaluative adjectives are subjective, e.g. challenging.

Example

The association between the presence/continuity of lignin and transpiration rate was assessed using the data of individual sweet 1 potato roots irrespective of the cultivar (Table 4). The distribution of the levels of transpiration rate was divided into three categories (as low, intermediate and high) and lignification was divided into two categories, according to the completeness of lignification. Pearson χ^2-tests indicated that lignification was significantly associated with lower transpiration rates at 6, 8, 10 and 13 days after wounding. The only day when no association with the transpiration rate was found was day 3, when presumably the wound healing process was not completed. We have previously postulated that water loss is the main cause of deterioration for roots during marketing in the tropics (Rees et al., 2003). This suggests that the L.S. for a cultivar could provide an important indication of the potential shelf life of that cultivar during marketing. Examples of cultivars with both high lignification scores and high shelf life were Yanshu 1 and, while SPK004 and Kemb10 have poor lignification scores and high weight loss (more than 20% after 10 weeks) (data not presented).

Evaluative adjectives in this example include: main, important, poor, high.

Task 1 Find the evaluative adjectives.

We found a relationship between root dry matter (DM) content and L.S. at lower humidities. The average root DM content for each cultivar is indicated in Fig. 2 and is significantly negatively correlated with the cultivar L.S. measured at 58% RH ($p = 0.027$) and nearly significant at 65% RH ($p = 0.073$) but not at 97%. Fig. 5a–Fig. 5d present the linear regression analysis for dry matter content and lignification score for four additional trials. In all cases, there was a negative correlation between dry matter and lignification score (p values 0.005, 0.065, 0.052 and 0.008).

Evaluative adjectives: ______________________________

Task 2 Match words 1)–8) to their meanings a–h.

1) accurate	a. displaying opposition
2) feasible	b. conforming exactly to a fact or to a standard
3) unbelievable	c. being worth doing
4) rewarding	d. feeling that something good will happen
5) satisfying	e. likely to be achieved and possible
6) positive	f. emphasizing how good, bad or extreme something is
7) significant	g. giving pleasure because it provides something you need
8) negative	h. large or important enough to have an effect

Task 3 Fill in the blanks with the evaluative adjectives given in the box.

Weak	significant	detectable	limited	strongest	strong

The phenol concentration of the UG solutions significantly affected the intensity of the target sensations (Fig. 3A and Table S2). According to the F values, the increase in phenol concentration had the 1) ________ effect on sourness, while it influenced the other target sensations much less. 2) ________ increases in intensity were observed in the samples with phenols from the UG extract compared to the sample without added phenol (0.00 g/L). Sourness increased from 3) ________ to 4) ________ across the phenol concentration range. Bitterness, astringency and saltiness showed 5) ________ increases in intensity, from barely 6) ________ to weak.

Unit 7 Conclusion

Main Contents	Learning Objectives
Subgenre: *Conclusion*	★ Understanding the structure of a Conclusion. ★ Identifying the moves of a Conclusion.
Organization: *General-specific*	★ Understanding the development of a text from general to specific.
Rhetorical functions: *Evidence and Conclusion*	★ Understanding evidence and conclusion. ★ Identifying the evidence development. ★ Understanding the relationship between evidence and conclusion.
Register: *Nominalization*	★ Distinguishing the types of nominalization. ★ Identifying the nominalization of verbs and adjectives, and figure out the functions of nominalization in a sentence.
Cohesion (6): *Lexical Chains*	★ Understanding the lexical chains. ★ Identifying the lexical chains. ★ Getting a general idea of a text under the aid of lexical chains.
Assignments: *Reading Comprehension and Vocabulary*	★ Reading: Understanding the Conclusion section better through in-depth reading. ★ Vocabulary: Learning to use the nouns derived from verbs or adjectives in a context.

7A Subgenre: *Conclusion*

Conclusion, as a part of a scientific paper, could be a separate part itself, while sometimes it is included in the Discussion part. Conclusion briefly reviews the research process, analyzes the result, and finally presents the research direction or prospect in one or two sentences. Thus, Conclusion requires rigorous logic and concise language. It is suggested that the research process, result and discussion should be briefly reviewed, since the core content has been thoroughly analyzed in the above parts of the article.

Significance, implication or application of the research is expected in Conclusion, which is the ultimate aim of the research. Conclusion may cover limitations of the research. Limitations are usually objective factors, such as small size of the sample, research devices or environment, instead of any subjective factors. Too many limitations may weaken the value of the research. Limitations of the research may introduce the unsettled problem of the topic or the research orientation of the subject, proposing the future direction of the research.

Hedging is often used in the Conclusion, since reasonable space is supposed to be left, especially in the social sciences.

There is no fixed length for the Conclusion, and it is usually short and concise.

1. The structure of a Conclusion

The Conclusion generally covers the following moves, some of which are obligatory, while some are optional.

Table 7.1 The Structure of a Conclusion

Move 1	briefly summarizing the research process, method and findings	optional
Move 2	answering research questions, explaining the research findings and analyzing the implications of the findings, or drawing a conclusion about whether the hypothesis is confirmed	obligatory

(To be continued)

(Continued)

Move 3	claiming the significance, values, or applications of the research	optional
Move 4	confessing limitations of the research	optional
Move 5	pointing out future research directions and prospects	optional

Example 1

Title of the paper: *Tobacco Rotated with Rapeseed for Soil-borne Phytophthora Pathogen Biocontrol: Mediated by Rapeseed Root Exudates*

[Briefly summarizing research findings] Rapeseed rotated with tobacco can effectively suppress black shank disease of tobacco caused by P. parasitica var. nicotianae in the field. **[Briefly explaining research findings]** Rapeseed roots can attract zoospores into the rhizosphere and then secrete a series of antimicrobial substances to kill them, which eventually caused P. parasitica var. nicotianae to lose its ability to spread or survive in the soil.

Example 2

Title of the paper: *Comprehensive Analysis of Wintersweet Flower Reveals Key Structural Genes Involved in Flavonoid Biosynthetic Pathway*

[Research process, method and results] In this study, a total of five flavonoids were identified in wintersweet flower. The transcriptome and proteome analysis yielded a large collection of genes encoding key enzymes involved in flavonoid biosynthesis. The quantitative DGE and q PCR analysis of the different developmental stages and the differently-colored varieties provided insights into the molecular mechanisms underlying the dramatic color changes in wintersweet. **[Explanation of the results]** These results demonstrated that ANS functions as an enzymatic on-off switch for anthocyanin biosynthesis and its expression is positively correlated with the expression of CHS, F3H1, FLS1, F3′H1, and UFGT genes. This expression profiling demonstrates that no gene products compete for common substrates to redirect the metabolic flux. **[Research significance]** Our results provide valuable insights into the molecular basis for high flavonoid content. Documenting the expression profiles of flavonoid synthases in the red and yellow varieties might facilitate the design of transgenics for engineering the flavonoid pathway and lay the foundation for future targeted-breeding approaches.

Example 3

Title of the paper: *Comparative Peptidomic Profile and Bioactivities of Cooked Beef, Pork, Chicken and Turkey Meat After in Vitro Gastro-Intestinal Digestion*

[Briefly summarizing research process, method and findings] In the present study, we applied an integrated approach combining peptidomic techniques with in vitro bioactivity assays. The four different meats were subjected to the harmonized INFOGEST in vitro gastrointestinal digestion protocol. Our study indicated that meat not only delivers important nutrients to humans, but also provides a source of bioactive peptides such as antioxidant as well as ACE- and DPP-IV-inhibitory peptides.

[Explaining research findings, analyzing the implications of the findings, and drawing a conclusion] Despite the limited differences in protein digestibility between the four types of tested meats, we found distinction in the peptidomic profiles after digestion. This discrepancy reflects the intrinsic differences in meat protein sequences. Moreover, these differences may result in the variation of the biological activities among species after in vitro digestion. Pork and turkey meats appeared to be the best sources of antioxidant peptides. Pork was also found to be the best source of DPP-IV-inhibitory peptides whereas chicken supplied peptides with the highest ACE-inhibitory activity.

[Claiming the limitation of the research] Different cooking temperatures and muscle types may led to relevant differences in peptide composition and abundance after in vitro gastrointestinal digestion. Such quantitative and qualitative differences may have an important impact on the release of bioactive peptides and related bioactivities of digested meat. Therefore, our results did not allow general conclusions to be drawn and further studies about the effect of cooking parameters and muscle types are warranted.

[Claiming the significance, values of the research] However, the present study lays the groundwork to discern meat from different species in the wake of their potential biological activities and bioactive peptides' profile after in vitro digestion. Indeed, this study aims to revise the concept of meat consumption, giving a new positive perspective, which has never been considered until now. **[Describing the necessity of further research]** However, more investigations and, especially, *in vivo* trials are needed to confirm the physiological significance of our observations.

Task 1 **Read the following Conclusions, match the sentences with the conclusion moves, and give a description of their functions. The underlined words may help.**

Conclusion 1

① This research provides novel information about *Scymnus* species and their relationships with the aphid species found in citrus crops in the Mediterranean basin. ② Our hypothesis on the quality of the diet influencing both predator' development and reproductive parameters, and consequently their field abundance and field distribution, was confirmed. ③ Even though the mixed diet that included *A. gossypii* and *A. spiraecola* was the most suitable for the development of both predator species, in the laboratory and in the field, *S. interruptus*, in particular, demonstrated better performance when preying on *A. gossypii*. ④ This may be one of the main factors determining the distribution and dynamics of the citrus aphid complex and its natural enemies in the Mediterranean basin.

Sentences	Move	Function

Conclusion 2

① Based on the data obtained in this current study, the use of New Zealand SM or CM can overcome the negative effects of a Ca and P restricted/deficient diet. ② It was found that SM was able to prevent the effects of consuming a diet restricted in Ca and P equally compared to CM, but the effect was achieved at a much lower intake of SM. ③ Although significant differences in organ distribution were observed with respect to selected macro and trace minerals related to milk type (specifically Zn, Cu, Co, and Fe), the data generated in the current study do not indicate a unique interaction between SM consumption and mineral absorption, for either macro or trace minerals. ④ With respect to non-essential mineral accumulation, this was found to be intricately linked to the intake of each respective mineral.

Sentences	Move	Function

Task 2 Read a Conclusion which is wrongly structured, restructure it by writing the numbers in the correct order, and complete the form that follows. The underlined parts may be helpful.

Title of the paper: *Impact of Biochar Amendment on the Uptake, Fate and Bioavailability of Pharmaceuticals in Soil-Radish Systems*

① Nonetheless, previous studies typically found similar behaviors of plant uptake and accumulation by various leafy and root vegetables [37, 49, 53], suggesting that the findings of this study may be broadly applicable to other types of vegetables.

② This showed that the impact on the plant uptake of pharmaceuticals by biochar amendment collectively depends on their sorption and dissipation in soils, as mediated by the presence of biochar.

③ Interestingly, the lincomycin concentration in radish was significantly increased by up to 48.2% in the soil amended with 1% biochar, mainly due to the increased persistence and pore water concentration of lincomycin in the biochar-amended soil.

④ This study provides evidence that biochar amendment of soils has a promising potential to simultaneously mitigate the accumulation of multiple pharmaceuticals in food crops and the associated human exposure.

⑤ Our results showed that when a sandy loam was amended with a pinewood biochar, especially at the rate of 1%, the uptake of 11 out of 15 studied pharmaceuticals by radish could be significantly reduced by up to 83.0% compared with those in the unamended soil.

⑥ Additionally, this study is limited by using only one root vegetable (radish).

Order	Move	Function
⑤	1	description of findings
	2	
		further explanation
		significance
	4	
		application

2. Sentence structures frequently used in a Conclusion

In each move of a Conclusion, there are some frequently used sentence structures, which are helpful in writing a Conclusion or identifying the elements of a Conclusion.

Table 7.2 Sentence Structures Frequently Used in a Conclusion

Moves	Sentence structures
Move 1 (Briefly summarize the research process, method and findings)	We applied an integrated approach combining ... with ... In this study, ... were identified. In this study, we successfully established ... model of ... Our results/study/research showed that ... It is evident from the study that ... Based on the data obtained in this current study ...
Move 2 (Answer research questions, explain the research findings and analyze the implications of the findings, or draw a conclusion about whether the hypothesis is confirmed)	This project was undertaken to design ... and evaluate ... The present study was designed to determine the effect of ... This study set out to assess the feasibility of ... This study set out to test the hypothesis that ... Our study indicated that ... This discrepancy/demonstration/classification/comparison/contrast/illustration reflects ...

(To be continued)

(Continued)

Moves	Sentence structures
	These differences may result in ... These results demonstrated that ... is positively correlated with ... This model bears importantly implications for the future development of ... The findings of this investigation complement those of earlier studies. These findings have significant implications for the understanding of how ... This study has raised important questions about the nature of ... Our hypothesis on ... was confirmed.
Move 3 (Claim the significance, values or applications of the research)	This (refers to the above findings) showed that ... This study may be broadly applicable to ... The result obtained from this work could be very valuable in decision making for ... The analysis/research/study provided insights into ... The research/study might facilitate ... The research/study may lay the foundation for ...
Move 4 (Confess limitations of the research)	This study is limited by ... However, further work is necessary to enrich our understanding of ... The results may not be generalized to ... We focus on only ... We have to point out that we do not ...
Move 5 (Point out the future research directions and prospects)	More investigations are needed to confirm ... Further studies about ... are warranted. However, ... remains to be further observed/investigated/analyzed/studied. It is expected that further analysis of ... will be significant in ...

Task 3 Read the above Examples 2&3 again and find out the language used to express moves which can help you to identify the elements.

Elements	Example 2	Example 3
Research process, methods and findings		
Explanations and/or implications		
Significance, values, and/or applications		
Limitations		
Future research direction and prospect		

7B Organization: *General-specific (GS)*

The general-specific (GS) pattern is widely found in the Conclusion part, largely because the author is inclined to present the findings of a study directly at the very beginning, after which the explanation, evaluation, supporting evidence and effects or results are presented. Thus, GS organization is often mingled with different functions aforementioned. GS pattern is often employed flexibly, especially when the findings are productive, and need to be explained or applied from different perspectives.

Example 1

Title of the paper: *Leachate Water Quality of Soils Amended with Different Swine Manure-based Amendments*

① Four forms of swine manure-based soil amendments (raw, compost, hydrochar, pyrochar) were evaluated for their effect on the soil fertility and leachate water quality characteristics in sandy soil. ② All amendments provided organic C and increased the soil CEC. ③ Because raw swine manure contains high concentrations of nutrients such as N, P, K, all four forms of amendments substantially increased the nutrient content of the soil. ④ Analysis of leachate following soil incubations showed that hydrochar amended soil released much lower concentrations of soluble N, P and K. ⑤ Pyrochar amended soil released over ten times more K, four times more P and twice the N concentration than hydrochar. ⑥ Even the control soil released more N than the hydrochar amended soil. ⑦ It appeared that the hydrochar increased both CEC and AEC of the soil, but subsequent testing of K, N adsorption isotherms and surface analysis via XPS suggested that these nutrients were not accumulated on the hydrochar surface. ⑧ It is not clear at this time how these were retained in the soil incubations, but the complex surface functionality of the hydrochar might interact with soil and produce conditions conducive to nutrients retention. Although it is still not clear how these nutrients were retained in the soil amended with hydrochar, it suggests a great potential for hydrochar as an alternative manure management option as the char can be applied to land nearby at high dose (in this case 2%) without creating potential environmental pollution problems from the leaching of high nutrient concentrations into water bodies.

In this Conclusion, we can see the general-specific pattern:

Sentence ① gives general and brief introduction of the study content (the effect of four swine manure-based soil amendments on the soil fertility and leachate water quality characteristics in sandy soil).

Sentences ②&③ present the overall findings (all four forms of amendments substantially increased the nutrient content of the soil).

Sentences ④–⑥ specifically explain the advantages of hydrochar.

Sentence ⑦ gives a further explanation of hydrochar and indicates the unclear aspect of the findings.

Sentence ⑧ introduces the application of hydrochar.

This Conclusion follows the general-specific pattern from both organization and content. The first three sentences gave a general presentation of the findings, which is followed by the specific illustration of the advantages of one specific amendment (hydrochar) over pyrochar amendment and even the control soil. Despite of the unclear reasons behind the fact that how these nutrients were retained in the soil incubations, this study suggests a great potential for hydrochar as an alternative manure management option.

Example 2

Title of the paper: *Water Content Variations and Pepper Water-use Efficiency of Yunnan Laterite Under Root-zone Micro-irrigation*

① During crop (pepper) cultivation, the Yunnan red loam soil had the most stable water content under root-zone micro-irrigation (IG), whereas the Yunnan red loam-yellow sand mixed soil had the smallest water content variations under surface-drip irrigation (IM). ② The pepper average water-use efficiency (WUE) under root-zone micro-irrigation was about 20.2% higher than that under surface-drip irrigation in the Yunnan red loam soil. ③ So, for Yunnan red loam soil, root-zone micro-irrigation has higher water saving efficiency compared to surface-drip irrigation. ④ While in the yellow sand soil, the pepper average water-use efficiency under root-zone micro-irrigation was about 16.6% lower than that under surface-drip irrigation. ⑤ Therefore, for yellow sand soil, surface-drip irrigation was more suitable. ⑥ The comparison of yellow sand soil, Yunnan red loam-yellow sand mixed soil and Yunnan red loam soil showed that water-use efficiency increases as the soil clay content increases.

The general-specific pattern is employed in this Conclusion:

Sentence ① states the finding.

Sentences ②&③ elaborate the reason why the root-zone micro-irrigation is more suitable for the Yunnan red loam soil.

Sentences ④&⑤ elaborate the reason why the surface-drip irrigation is more suitable for the yellow sand soil.

Sentences ⑥ makes a broad statement of comparison.

In the Conclusion section, general-specific pattern is convenient and efficient to introduce the following necessary explanation, evaluation, effect, comparison, etc. However, popular as the general-specific pattern is, the specific-general pattern can also be found in the Conclusion section. In the specific-general pattern, the specific explanation or illustration is followed by a general conclusion.

Task 1 Read the following Conclusion and finish the exercises.

① The IIFI was found to be a suitable approach for describing individual feeding behavior patterns. ② This index has the ability to integrate most of the information on feeding behavior obtained using several conventional feeding behavior variables. ③ Additionally, IIFI takes into account intra-animal variability across days because it includes the daily distribution of feeding events over a period of one week. ④ Pigs with a large IIFI value show a less regular pattern of meals and feed intake. ⑤ Although significant associations were found between IIFI and the number of daily meals, meal duration and FI per meal, pigs with similar values for the conventional variables could have different IIFI values because of differences in the dynamics of feed consumption across days. ⑥ In comparison with the conventional variables, the IIFI was more suitable for identifying differences under different nutritional strategies, which indicates the potential that this index offers for better describing the feeding behavior of pigs. ⑦ This index also has the potential of being used to predict the existence of sanitary challenges in livestock as those challenges interfere in the feeding behavior at the level of visits.

1) This Conclusion begins with:
   A statement of definition □
   A statement of finding □
   A statement of specific information □
2) This Conclusion is organized in the pattern of:
   General-specific □
   Specific-general □
3) The following steps describe the flow of information. Read the Conclusion again and complete the table by writing down the sentence number for each step.

Step 1	Generally state the finding.	Sentence ______
Step 2	Explain why IIFI is a suitable approach.	Sentence ______
Step 3	Explain how IIFI value reflects feeding behavior.	Sentence ______
Step 4	Indicate the limitation of IIFI.	Sentence ______
Step 5	Describe the advantages of IIFI over conventional variables.	Sentence ______
Step 6	Describe other potential application of IIFI.	Sentence ______

Task 2 Read the following GS text and finish the exercises.

① It has been shown in this study that soil ecosystems are associated with factors that influence their composition. ② In addition, this work has confirmed that heavy metal concentration represents another key factor. ③ Heavy metals play an important but complex role in soil microbial communities by accelerating or inhibiting decomposition of polymers and organic compounds. ④ Heavy metals can be aggregated into minerals, or incorporated into basic microbial secondary metabolites secreted by microbes that interact with metal surfaces (Bahadur et al., 2016). ⑤ However, the exact impact of Au and Pt on the soil microbial community is difficult to predict given the array of biotic and abiotic factors unique to a soil such as the degree of adaptation of the microbial community and vegetation (biotic) together with soil organic matter content, moisture and pH (abiotic). ⑥ Amendment with increasing concentration of mobile Pt or Au complexes caused reductions in soil respiration rates in most tested soils, demonstrating that both Au and Pt complexes are toxic to soil microbial communities. ⑦ However, chemical analyses and specification studies would be needed to further investigate if ions or complexes cause toxicity. ⑧ These reactions were related to soil type and concentrations of the amendment. ⑨ While mobile Au appears to be somewhat more toxic to soil microbial communities, toxicity levels appear to be linked to soil properties.

1) The general statement is: ________________________________

2) This paragraph serves as the Conclusion section in the paper. Can you find anything special about general-specific pattern in this paragraph? Do you think it is acceptable?

7C Rhetorical functions: *Evidence and Conclusions*

Rhetorical functions in paragraphs can be divided into three main macro-functions, namely, *describe, explain* and *persuade*. The function of *persuade* generally consists of problem and solution, argument claim and support/counter claim as well as evidence and conclusions. Featured as drawing a conclusion with sufficient supporting evidence, evidence and conclusions often function as the persuasion of the writer's position, result affirmation, hypothesis justification and so on. Evidence may include facts, data, arguments or claims reasoned by the writer or cited from others.

Example

The title of the paper: *Dose-related Changes in Respiration and Enzymatic Activities in Soils Amended with Mobile Platinum and Gold*

[Augument claim] It has been shown in this study that soil ecosystems are associated with factors that influence their composition. In addition, this work has confirmed that heavy metal concentration represents another key factor. **[Evidence]** Heavy metals play an important but complex role in soil microbial communities by accelerating or inhibiting decomposition of polymers and organic compounds. Heavy metals can be aggregated into minerals, or incorporated into basic microbial secondary metabolites secreted by microbes that interact with metal surfaces (Bahadur et al., 2016). **[Counter Claim]** However, the exact impact of Au and Pt on the soil microbial community is difficult to predict given the array of biotic and abiotic factors unique to a soil such as the degree of adaptation of the microbial community and vegetation (biotic) together with soil organic matter content, moisture and pH (abiotic). **[Evidence]** Amendment with increasing concentration of mobile Pt or Au complexes caused reductions in soil respiration rates in most tested soils, demonstrating that both Au and Pt complexes are toxic to soil microbial communities. **[Counter Claim]** However, chemical analyses and specification studies would be needed to further investigate if ions or complexes cause toxicity. **[Conclusion]** These reactions were related to soil type and concentrations of the amendment. While mobile Au appears to be somewhat more toxic to soil microbial communities, toxicity levels appear to be linked to soil properties.

From the annotations and lines in the text, functions are most transferable across academic texts. Identifying functions is a way to catch author's idea flow, to recognize relationships between ideas and how these change through a text. Awareness of functions helps us to understand the meaning of linking expressions, e.g. *however*, *in addition*, *moreover*.

Task 1 **Read the following Discussion part, figure out specific functions, and complete the form that follows.**

The title of the paper: *Community Food Environment Moderates Association Between Health Care Provider's Advice on Losing Weight and Eating Behaviors*

① This study examined if the community food environment moderates the relationship between receiving weight loss advice from a HCP and consumption of food and beverages in an OW/OB population. ② Interaction and stratified analyses revealed that receiving HCP's advice on losing weight was associated with a lower reported consumption of total SSB, soda, and sweetened fruit drinks when participants lived near a small grocery store, or far from a supermarket, LSR, or convenience store. ③ However, when participants lived near supermarkets, LSRs, or convenience stores, there was no association between HCP's advice and reported SSB consumption. ④ We found no association with respect to fruit, vegetable, salad or fast-food consumption. ⑤ These results elucidate the complex role of context (i.e. community food environment) on the effect of HCP's weight loss advice.

Function	Sentence number
	①
Evidence 1	

(To be continued)

(Continued)

Function	Sentence number
	③
	④
Conclusion	

Task 2 **Read the following Discussion part, figure out specific functions, and complete the table that follows.**

The title of the paper: *Effects of Micronutrient Fertilization on the Overall Quality of Raw and Minimally Processed Potatoes*

① This study has highlighted a significant effect of foliar micronutrient application on physico-chemical and nutritional characteristics of raw potatoes, through an increase of firmness, total solids content, total soluble solids content, reducing sugars and ascorbic acid content. ② Improved quality characteristics of Micro+ raw tubers allowed better performance of Micro+ minimally processed potatoes during storage time. ③ In particular, Micro+ minimally processed potatoes compared to Micro– ones showed higher firmness, total solids, total soluble solids, reducing sugars and ascorbic acid content which contributed to a less propensity to undergo browning during storage. ④ At the same time, Micro+ fertilized samples compared to Micro– showed lower microbial populations (mesophilic, psychrotrophic and lactic bacteria), which remained below 8 log CFU/mL up to the end of cold storage, allowing three days longer shelf life. ⑤ In conclusion, micronutrient fertilization could be a valuable pre-harvest treatment as it proved effective at improving nutritional quality of both raw tubers and minimally processed potatoes and at extending their shelf life. ⑥ However, to have a more complete picture on the effects of micronutrient fertilization on overall quality, it is necessary to define the mineral profile of the samples, since certain micronutrients are essential for a healthy diet. ⑦ In addition, given that potatoes are always cooked before consumption, the sensory, including taste, and physicochemical characteristics of both raw and minimally processed tubers after home preparation, also need evaluating.

Function	Main idea	Sentence number
Introduction	A review of what this study has found.	①
Evidence 1		
Evidence 2	Micro+ fertilized samples showed lower microbial populations in comparison with Micro− ones, allowing longer shelf life.	④
Conclusion		⑤

7D Register: *Nominalization*

Up till now, you may have found and perceived such language features of academic language as conciseness, preciseness, objectivity and formality. From the perspective of register, which refers to the use of different styles of language in different situations, the style of academic texts is different from that of others in various ways, one of which lies in nominalization, for its unique empowerment of verbs or adjectives transformation to function as nouns. Nominalization refers to the process that the meaning of a verb or an adjective is expressed by a noun (generally speaking, the noun derived from the verb or adjective) or a noun phrase, during which the syntax structure need to be transformed accordingly, aiming to achieve formal, concise and condensed register in academic writing. Sometimes, the noun and the verb of a specific meaning share the same word, such as *change*, *switch*.

1. Nominalization of verbs

To meet the demands of presenting a lot of information in a clear and unambiguous way, many verbs may be transformed into nouns with condensed information, which reflects the formality and conciseness of academic writing. For example:

1a. Flower color has successfully elucidated the fundamental principles of genetics.

1b. Flower color has successfully contributed to the elucidation of the fundamental principles of genetics.

2a. To date, no study has shown that PERV (内源性逆转录病毒) could be transmitted to humans in a clinical setting.

2b. To date, no study has shown PERV transmission to humans in a clinical setting.

3a. Porcine organs have not been used clinically yet, because immune system of humans is incompatible with that of pigs, and PERV may be transmitted to humans.

3b. The clinical use of porcine organs has been hindered by immunological incompatibilities (2) and by the potential risk of PERV transmission.

In 1b and 2b, the verbs (*elucidate, transmit*) are transformed into noun derivations respectively. Compared with 1a, 1b is apparently more formal. 2b is more concise than 2a. In 3b, the information of two sentences in 3a is condensed into 2 nouns (*incompatibility and transmission*).

The following paragraph is a Conclusion section of a paper. The underlined words are nominalized from verbs, which function as subject, object and adverbial modifier in sentences.

Example 1

Title of the paper: *Comprehensive Analysis of Wintersweet FLower Reveals Key Structural Genes Involved in FLavonoid Biosynthetic Pathway*

In this study, a total of five flavonoids were identified in wintersweet flower. The transcriptome and proteome **analysis** yielded a large collection of genes encoding key enzymes involved in flavonoid **biosynthesis**. The quantitative DGE and q PCR **analysis** of the different developmental stages and the differently-colored **varieties** provided insights into the molecular mechanisms underlying the dramatic color changes in wintersweet. These results demonstrated that ANS functions as an enzymatic on-off switch for anthocyanin biosynthesis and its **expression** is positively correlated with the **expression** of CHS, F3H1, FLS1, F3′H1, and UFGT genes. **This expression** profiling demonstrates that no gene products compete for common substrates to redirect the metabolic **flux**. Our results provide valuable insights into the molecular basis for high flavonoid content. Documenting the expression profiles of flavonoid synthases in the red and yellow varieties might facilitate the design of transgenics for engineering the flavonoid pathway and lay the foundation for future targeted-breeding approaches.

Different sentence elements	Examples
Subject (underlined with double line)	analysis; varieties; expression
Object (underlined with wave)	expression; flux
Adverbial modifier (underlined with dotted line)	biosynthesis

2. Nominalization of adjectives

From the perspective of grammar, adjectives can function as attributive in a sentence. However, if an adjective can be transformed into a noun, which often required by grammar rules and sentence organization, it could function as different sentence elements in grammar flexibly, while retaining the same meaning as the original adjective. For example,

1a. The immune system of humans is incompatible with that of pigs.

1b. immunological incompatibilities

2a. Flower color is very important for plant ecology and evolution.

2b. Flower color is of paramount importance for plant ecology and evolution.

In 1a, the information of a sentence is condensed into 2 words in 1b. Compared with the adjective *incompatible*, its corresponding noun, *incompatibility*, allows for more flexible organization of a complex sentence, and meets the needs of the academic writing — formality and conciseness. As you can see, 2b is also more formal than 2a.

More examples of nominalization of adjectives can be found in the following paragraph. The grammatical function of nouns is more powerful than that of adjectives, so an adjective, together with its nominalized form, can function as many more sentence elements.

Example 2

Title of the paper: *Comparative Peptidomic Profile and Bioactivities of Cooked Beef, Pork, Chicken and Turkey Meat After in Vitro Gastro-intestinal Digestion*

Despite the limited **differences** in protein **digestibility** between the four types of tested meats, we found **distinction** in the peptidomic profiles after digestion. This **discrepancy** reflects the intrinsic **differences** in meat protein sequences. Moreover, these **differences** may result in the **variation** of the biological activities among species after in vitro digestion. Pork and turkey meats appeared to be the best sources of antioxidant peptides. Pork was also found to be the best source of DPP-IV inhibitory peptides whereas chicken supplied peptides with the highest ACE-inhibitory activity.

Different cooking temperatures and muscle types may led to relevant

differences in peptide composition and **abundance** after *in vitro* gastrointestinal digestion. Such quantitative and qualitative **differences** may have an important impact on the release of bioactive peptides and related bioactivities of digested meat. Therefore, our results did not allow general conclusions to be drawn and further studies about the effect of cooking parameters and muscle types are warranted.

Different sentence elements	Examples
Subject (underlined with double line)	discrepancy; differences; differences
Object (underlined with curve)	distinction; differences; variation; differences; abundance
Adverbial modifier (underlined with dotted line)	differences; digestibility

The underlined words in bold are nominalized adjectives that function as different sentence elements. For example, in the first sentence, *digestibility*, derived from *digestible*, is the adverbial modifier of *differences*, which is itself a nominalization of *different*. The word *distinction*, derived from *distinct/distinctive*, functions as the object in the second sentence.

From the perspective of cohesion, nominalization can also function as a cohesive tie to link the former information with the latter information. In the above example, *this discrepancy* in the second sentence summarizes the information in the first sentence while introducing the following information. The noun phrase, *these differences*, in the third sentence serves the similar purpose.

Task **Read the following sentences and identify the nominalized words, their types of nominalization and the sentence constituents.**

1) Our hypothesis on the quality of the diet influencing both predators' development and reproductive parameters, and consequently their field abundance and field distribution, was confirmed.

2) This may be one of the main factors determining the distribution and dynamics of the citrus aphid complex and its natural enemies in the Mediterranean basin.
3) Although significant differences in organ distribution were observed with respect to selected macro and trace minerals related to milk type (specifically Zn, Cu, Co, and Fe), the data generated in the current study do not indicate a unique interaction between SM consumption and mineral absorption, for either macro or trace minerals.
4) With respect to non-essential mineral accumulation, this was found to be intricately linked to the intake of each respective mineral.
5) This study provides evidence that biochar amendment of soils has a promising potential to simultaneously mitigate the accumulation of multiple pharmaceuticals in food crops and the associated human exposure.

7E Cohesion (6): *Lexical Chains*

Lexical chains are connected with theme-related words, which lies in the paper as a whole, in each section of the paper as well as in each paragraph of the sections. Lexical chains, which are especially apparent in abstract, run through the whole paper and construct the field of a discourse. Generally speaking, the content words in the title and the key words of the paper are contained in lexical chains. Lexical chains are developed by repetition, deviation or classification of the theme-related words.

Example 1

Title of the paper: *The Effect of the Method of Plant Protection on the Quality of Remontant Strawberry Cultivars Grown in a Gutter System Under Covers*

Fruit measurements and chemical analyses were performed on a random sample for each combination of 40 fruits. Strawberry fruit firmness [N] was measured with a TA 500 Lloyd Texture Analyzer using a 6.35 mm diameter tip. The measurement of fruit firmness was performed 1 for one fruit. Soluble solids content SSC (%) and total acidity TA (% citric acid) were determined in the juice of strawberries, whose firmness was previously measured using an Atago Pal-BX/Acid 4 instrument. The soluble solids content to total acidity ratio (SSC/TA) was calculated. The fruit respiration rate (mg CO_2 kg^{-1} h^{-1}) was measured (on a sample of 9 strawberries from the combination) with an Air Tech 2500-P CO_2 analyzer.

The first paragraph, extracted from the Discussion part, explains one of the research methods — fruit quality measurement in the study. The words underlined with single lines formed a lexical chain about the topic. The *strawberry fruit firmness, soluble solids* *content SSC (%), total acidity TA (% citric acid), The fruit respiration rate (mg CO_2 kg^{-1} h^{-1})* explain the data that need to be measured and analyzed in this study, and the words underlined with waves, namely, *a TA 500 Lloyd Texture Analyzer using a 6.35 mm diameter tip, an Atago Pal-BX/Acid 4 instrument* and *an Air Tech 2500-P CO_2 analyzer* state the instruments been used to measure and analyze the data of each item in the study, which also constitute the lexical chain of this paragraph.

Probably, more than one lexical chain can be identified in a paper, and the interaction and intersection among lexical chains not only achieves coherence and cohesion of the discourse, but also helps strengthen theme presentation and argument development. The following example is extracted from the discussion section. The first sentence is an introduction of the topic, and the lexical chain begins from the second sentence.

Example 2

Title of the paper: *Climate Change Might Lead to Substantial Niche Displacement in One of the Most Biodiverse Regions in the World*

Our analysis suggests that the climate space of the Atlantic Forest (AF) will get hotter and drier in the next 20–80 years. In addition, 10%–35% of the currently climatic space will shift. These findings are in line with the latest IPCC assessment report for South America (IPCC 2022), which suggested that temperature in all South America subregions will most likely increase at rates greater than the global average. Precipitation, however, has been predicted to be more variable and regionally specific. While there is an expected decrease in precipitation in the harsher zone (i.e. the northern part of the AF), the milder zone (i.e. the southern part of the AF) will most likely experience an increase in precipitation according to IPCC predictions (IPCC 2022). In addition, the intensity and frequency of extreme precipitation and floods are projected to increase in the southern parts of the AF, whilst the northern parts will likely experience an increase in intensity of droughts (IPCC 2022).

The underlined words formed a lexical chain of *climate space*. From this lexical chain we can easily understand what *climate space* refers to in this paper. Another key word of this paper is *climate change*, which forms another lexical chain. Read the paragraph again, and pay attention to the underlined words.

Our analysis suggests that the climate space of the Atlantic Forest (AF) will get hotter and drier in the next 20–80 years. In addition, 10%–35% of the currently climatic space will shift. These findings are in line with the latest IPCC assessment report for South America (IPCC 2022), which suggested that temperature in all South America subregions will most likely increase at rates greater than the global average. Precipitation, however, has been predicted to be more variable and regionally specific. While there is an expected decrease in precipitation in the harsher zone (i.e. the northern part of the AF), the milder zone (i.e. the southern part of the AF) will most likely experience an increase in precipitation according to IPCC predictions (IPCC 2022). In addition, the intensity and frequency of extreme precipitation and floods are projected to increase in the southern parts of the AF, whilst the northern parts will likely experience an increase in intensity of droughts (IPCC 2022).

Through this lexical chain about the climate change, we can figure out that the main change that the researcher focuses on are *temperature* and *precipitation*, and the change trends include *increase* and *decrease*. From the lexical chain, we can get a general impression of the research content. In other words, lexical chain, which reflects the cohesion of the text, functions as the guide of the main idea of the paper.

Task 1 **Read the following paragraph, and identify the lexical chains in it. The bold part is the topic sentence of the paragraph.**

The crop protection methods used had little effect on marketable yield and fruit weight. This depended on the variety of climatic conditions for a given growing season. In contrast, chemical and biological protection significantly reduced the size of the non-commercial yield, a relationship that differed from year to year and for each variety tested. Chemical protection effectively reduced the incidence of the most dangerous fruit pathogens on the studied strawberry varieties throughout the research period. The effectiveness of biological preparations depended on the climatic conditions prevailing in each year of this study, and the susceptibility of the variety to a given pathogen also influenced the effectiveness of biological protection. During laboratory tests of fruit quality, a significant influence of the method of plant protection on the studied quality properties of strawberries was found. However, this influence manifested itself in different ways, depending on the cultivar and the method of protection.

Task 2 Figure out the lexical chains and other cohesive ties in the following paragraph.

This paragraph introduces six features or aspects of IPEC-J2, these aspects go around the topic IPEC-J2, so there is an upper lexical chain on a whole. As for every aspect, there are several lower lexical chains with overlaps (such as the topic: IPEC-J2).

Title: *Porcine IPEC-J2 Intestinal Epithelial Cells in Microbiological Investigations.*

IPEC-J2 cells are porcine intestinal columnar epithelial cells that were isolated from neonatal piglet mid-jejunum. This cell line forms polarized monolayers with high transepithelial electrical resistance when cultured on 0.4 μm pore-size filters. The cell line is unique in that it is derived from small intestinal tissue (compared to the common human colon-derived lines HT-29, T84, and Caco-2) and is not transformed (compared to the porcine small intestinal line, IPI-2I). Porcine intestinal epithelial cells more closely mimic human physiology than analogous rodent cell lines (e.g. IEC-6 or IEC-18), which is important in studies of zoonotic infections; in addition, they provide specificity to study porcine-derived infections. IPEC-J2 cells are increasingly being used in microbiological studies to examine the interactions of various animal and human pathogens, including Salmonella enterica and pathogenic Escherichia coli, with intestinal epithelial cells. The IPEC-J2 cell line has also been employed in some probiotic studies, in which the cells have been used as an initial screening tool for adhesiveness and anti-inflammatory properties of the potential probiotic microorganisms. The validity of these studies is not clear as follow-up studies to assess the efficacy of the probiotics *in vivo* have not been published to date. The aims of this review are to provide a comprehensive overview of the microbiological studies that have been conducted with IPEC-J2 cells and a reference guide of key cellular and immune markers that have been identified in this cell line that may prove to be useful in future studies.

Cohesive ties					
Grammatical cohesion			Lexical cohesion		
Conjunction	Substitution	Reference	Lexical reiteration: repetition, synonym, superordinate	Lexical chains	General nouns

7F Assignments: *Reading Comprehension and Vocabulary*

Part one Reading comprehension

Task 1 Read the following text and answer the questions that follow.

NUE 氮素利用效率（Nitrogen use of efficiency）

inorganic *adj.* 无机的

multi-nutrients *n.* 多养分肥料

optimum *adj.* 最优的

acidity *n.* 酸度

gradient boosted tree regression models 渐进梯度回归树模型（一种R语言机器算法模型，多用于处理分析不同类型的数据）

① A systematic analysis was done for 13 long-term experiment sites to quantify the effect of long-term fertilizer inputs, soil properties and climatic variables on the **NUE**. ② The NUE of cropland systems in southern China showed a high variation with values ranging from –6%–127%. ③ Lowest NUE was found in cases where crops were fertilized with **inorganic** N fertilizers only, and the NUE increased when **multi-nutrients** or organic manures were applied. ④ The soil, climatic and fertilizer inputs together explained 46%–85% of the variation in NUE. ⑤ The main variables controlling NUE across the sites were the pH, total P inputs, available P and K and the duration of the fertilizer regimes applied. ⑥ In line with our hypothesis, soil pH had an **optimum** pH around 6, with lower values being associated with reduced nutrient availability for P, Ca, Mg and K availability. ⑦ In addition to soil **acidity**, NUE increased with available soil P (AP) reaching a plateau at an AP near 50–100 mg/kg. ⑧ Using generalized linear regression modelling, we found that the NUE decreases with an increase in N input, in line with our hypothesis, but this effect was less evident when applying **gradient boosted tree regression models**. ⑨ Soil organic carbon had a positive impact on NUE, whereas the evidence was less clear for the impact of clay. ⑩ Sites with higher **precipitation** rates had lower NUE values, whereas NUE increased with temperature. ⑪ Furthermore, we found that the optimum NUE was found when 30%–40% of the N input is given as manure. ⑫ Using **empirical models** trained on data from the long-term experiments, we found that the NUE can increase from 30% to 42% up to 42%–67% by altering the soil nutrient levels and the N dose and fertilizer type. ⑬ Additional field evidence is needed to explore the full potential of fertilizer strategies to enhance NUE.

precipitation *n.* 降水

empirical model 经验模型

Questions

1) How many of these moves can you find in this Conclusion? Identify the sentences that correspond to each move in the table below.

Move	Typical elements	Sentence number
Move 1	briefly summarizing research process, method and findings	
Move 2	answering research questions, explaining research findings and analyzing the implication of the findings, or drawing a conclusion about whether the hypothesis is confirmed	
Move 3	claiming the significance, values, or applications of the research	
Move 4	confessing limitations of the research	
Move 5	pointing out future research direction and prospect	

2) How many findings are there in this Conclusion?
3) Whether the hypotheses in this study is confirmed and why?
4) What are the significant variables that have effects on the NUE?
5) Find out the lexical chain which described the *effect* in this Conclusion.

Part two Vocabulary

Nouns not only undertake most sentence elements, such as subjects or objects, but also carry condensed information themselves. To guarantee the formality and conciseness of academic text, nominalization is common in academic texts. So it is an indispensable ability to transform verbs or adjectives into nouns. Finish the following tasks to check whether you can write down the nouns derived from verbs or adjectives.

Task 1 Transform the following words into nouns.

Verbs	Nouns	Verbs	Nouns
define		identify	
distribute		imply	
indicate		interpret	
construct		evaluate	
investigate		contribute	

Adjectives	Nouns	Adjectives	Nouns
different		significant	
subsequent		equivalent	
mental		flexible	
accurate		identical	

Task 2 Fill in the blanks with the appropriate form of the given words.

1) Their cancers are not so clearly tied to radiation ________. (expose)
2) Market ________ also means searching for increased usage among present customers or going for a different market, such as senior citizens. (modify)
3) The optimum ________ conditions of shionone from herbs tatarian aster root were studied. (extract)
4) The resources at our ________ could have been better utilized. (dispose)
5) Strong productivity growth has been achieved partly through the ________ of many mid-skilled jobs. (eliminate)
6) The feedback may include the ________ of knowledge about your drugs potential side effects. (accumulate)

7) So how to explain the seeming ________ with 2 larger studies published only a year ago? (contradict)
8) Now both Republicans and Democrats are screaming about price ________ at the gas pump. (manipulate)
9) With dedication, ________ and ingenuity, we can put an end to the diabetes epidemic. (persistent)
10) During transcription, this molecule encodes and carries information from genes to sites of protein ________. (synthetic)

Appendix

Articles Used as Examples

Błaszczyk, J., et al. (2022). The Effect of the Method of Plant Protection on the Quality of Remontant Strawberry Cultivars Grown in a Gutter System under Covers. *Agriculture*, *12*(12), 2041.

Burrow, K. (2020). The Effect of Sheep and Cow Milk Supplementation of a Low Calcium Diet on the Distribution of Macro and Trace Minerals in the Organs of Weanling Rats. *Nutrients*, *12*, 594.

Chilton, M., et al. (2009). Food Insecurity and Risk of Poor Health Among US-Born Children of Immigrants. *American Journal of Public Health*, *99*(3). DOI: 10.2105/AJPH.2008.144394.

Davidson, A. (1999). *The Oxford Companion to Food.* Oxford: Oxford University press.

Elleby, C. (2020). Temporary Reduction in Daily Global CO_2 Emissions During the Virus Forced Confinement. *Environmental & Resource Economics*, *8*(4), 1–13.

Ermakova, I. (2005). Influence of Genetically Modified Soya on the Birth-weight and Survival of Rat Pups. http://www.oeko.de/oekodoc/277/2006-002-en.pdf [Accessed: 23 April 2022]

Escribanoa, S., et al. (2016). Impact of 1-methylcyclopropene Treatment on the Sensory Quality of 'Bartlett' Pear Fruit. *Postharvest Biology and Technology*, *111*, 305–313.

Fang, Y., et al. (2016). Tobacco Rotated with Rapeseed for Soil-borne Phytophthora Pathogen Biocontrol Mediated by Rapeseed Root Exudates. *Frontiers in Microbiology*, *7*, 894. DOI:10.3389/fmicb.2016.00894. https://doi.org/10.3389/fmicb.2016.00894

Gaudin, A.M., et al. (2015). Wheat Improves Nitrogen Use Efficiency of Maize and Soybean-based Cropping Systems. *Agriculture, Ecosystems & Environment.* https://xueshu.baidu.com/usercenter/paper/show?paperid=dcec6d992f577ab73c64d2e4c7b343c1&site=xueshu_se

Gómez-Ramírez, A., et al. (2017). Surface Chemistry and Germination Improvement of Quinoa Seeds Subjected to Plasma Activation. *Scientific Reports*, *7*, 5924.

Ierna, A., et al. (2017). Effects of Micronutrient Fertilization on the Overall Quality of Raw and Minimally Processed Potatoes. *Postharvest Biology and Technology*, *134*, 38–44. DOI: 10.1016/j.pmedr.2019.100926. https://doi.org/10.1016/j.pmedr.2019.100926

Izquierdo-Gomeza, D., et al. (2015). Effects of Coastal Fish Farms on Body Size and Isotope Composition of Wild Penaeid Prawn. *Fisheries Research*, *172*, 50–56.

Leprieur, F., et al. (2008). Fish Invasions in the World's River Systems: When Natural Processes

Are Blurred by Human Activities. *PLoS Biol*, *6*(2), e28. DOI:10.1371/journal.pbio.0060028 [Accessed: 25 July 2022]

Li, S., et al. (2017). Cinnamic, Myristic and Fumaric Acids in Tobacco Root Exudates Induce the Infection of Plants by Ralstonia solanacearum. *Plant and Soil*, *412*(1–2), 381–395.

Li, Y. B., et al. (2020). Impact of Biochar Amendment on the Uptake, Fate and Bioavailability of Pharmaceuticals in Soil-Radish Systems. *Journal of Hazardous Materials*, *398*, 122852.

Liu, X.J., et al. (2022). Dam Construction Impacts Fish Biodiversity in a Subtropical River Network, China. *Diversity*, *14*, 476.

Lorts, C., et al. (2019). Community Food Environment Moderates Association Between Health Care Provider's Advice on Losing Weight and Eating Behaviors. *Preventive Medicine Reports*, *15*, 100926.

López-Isasmendi, G., et al. (2019). Aphicidal Activity of Bacillus Amyloliquefaciens Strains in the Peach-potato Aphid (Myzus persicae*). Microbiological Research*, *226*, 41–47.

Makanga, U., et al. (2022). Obstacle-induced Lateral Dispersion and Nontrivial Trapping of Flexible Fibers Setting in a Viscous Fluid. *Fluid Dynamics*, *9*(21).

Martini, S., et al. (2019). Comparative Peptidomic Profile and Bioactivities of Cooked Beef, Pork, Chicken and Turkey Meat after in Vitro Gastro-intestinal Digestion. *Journal of Proteomics*, *208*. DOI: 10.1016/j.jprot.2019.103500. https://doi.org/10.1016/j.jprot.2019.103500

Mitiku, D. H., Abera, S., & Bussa, N. (2018). Evaluation of Physicochemical Properties and Sensory Attributes of Leavened Bread Produced from Composite Flours of Wheat (triticum aestivum l.) and Sweet Potato (ipomoea batatas l.). *Science, Technology and Arts Research Journal*, *5*(1), 95.

Murakami, M., et al. (2023). Climate Change Might Lead to Substantial Niche Displacement in One of the Most Biodiverse Regions in the World. *Plant Ecology*, *224*, 403–415.

Oirschot, Q. E. A. V., et al. (2006). Sweetpotato Cultivars Differ in Efficiency of Wound Healing. *Postharvest Biology and Technology*, *42*(1), 65–74.

Ortel, C., et al. (2021). Influence of Soybean Management Decisions on the Subsequent Rice Crop's Response to Nitrogen. *Agronomy Journal*, *113*.

Ro, K.S., et al. (2015). Leachate Water Quality of Soils Amended with Different Swine Manure-based Amendments. *Chemosphere*, *142*, 92–99. DOI:10.1016/j.chemosphere.2015.05.023. https://doi.org/10.1016/j.chemosphere.2015.05.023

Sadique, S., et al. (2021). Effect of Foliar Supplied PGRs on Flower Growth and Antioxidant Activity of African Marigold (Tagetes erecta L.). *Horticulturae*, *7*, 378.

Shar, S., et al. (2021) Dose-related Changes in Respiration and Enzymatic Activities in Soils Amended with Mobile Platinum and Gold. *Applied Soil Ecology*, *157*, 103727. DOI: 10.1016/j.apsoil.2020.103727. https://doi.org/10.1016/j.apsoil.2020.103727

Singh, K., Risse, M., Worley, J., Das, K. C., & Thompson, S. (2007, June 17–20). *Adding Value to Poultry Litter Using Fractionation, Pyrolysis, and Pelleting* [Conference presentation].

Minneapolis, Minnesota, United States. DOI: 10. 13031/2013.23096.

Sjöholm, E., et al. (2020). 3D-Printed Veterinary Dosage Forms — A Comparative Study of Three Semi-solid Extrusion 3D Printers. *Pharmaceutics*, *12*, 1239.

Sørensen, J. F. L. & Jørgensen, H. P. (2022). Rural Development Potential in the Bioeconomy in Developed Countries: The Case of Biogas Production in Denmark. *Sustainability*, *14*, 11077. https://doi.org/10.3390/su141711077

Song, W. J., et al. (2022). An Efficient Meshless Method for Bimaterial Interface Cracks in 2D Thin-layered Coating Structures. *Journal Pre-proof.* https://www.sciencedirect.com/science/article/abs/pii/S0893965922000994

Wen, J., et al. (2020). Effects of Probiotic Litchi Juice on Immunomodulatory Function and Gut Microbiota in Mice. *Food Research International*. *137*, 109433.

Wang, H., et al. (2019). Effects of Dietary Energy on Growth Performance, Rumen Fermentation and Bacterial Community, and Meat Quality of Holstein-Friesians Bulls Slaughtered at Different Ages. *Animals*, *9*(12). DOI：10.3390/ani9121123.

Xu, Y. (1999). Agricultural Productivity in China. *China Economic Review*, *10*(2).

Yerli, C., Sahin, U., Mehmet Kiziloglu, F., *et al.* (2022). Deficit Irrigation with Wastewater in Direct Sowed Silage Maize Reduces CO_2 Emissions from Soil by Providing Carbon Savings. *Journal of Water & Climate Change*, *13*(7).

Yang, N., et al. (2018). Comprehensive Analysis of Wintersweet Flower Reveals Key Structural Genes Involved in Flavonoid Biosynthetic Pathway. *Gene: An International Journal Focusing on Gene Cloning and Gene Structure and Function*, *676*.

Yi, D., et al. (2019). Physiochemical Properties of Rice with Contrasting Resistant Starch Content. *Journal of Cereal Science*, 89.

Zeinipour, M. (2018). Agroinfiltration: A Rapid and Reliable Method to Select Suitable Rose Cultivars for Blue Flower Production. *Physiol Mol Biol Plants*, *24*(3), 503–511.

Zhang, Y. J., et al. (2022). Water Content Variations and Pepper Water-use Efficiency of Yunnan Laterite Under Root-zone Micro-irrigation. *Frontiers in Plant Science*, *13*, 918288. DOI:10.3389/fpls.2022.918288. https://doi.org/10.3389/fpls.2022.918288

References

Alexander, O., Argent, S., & Spencer, J. (2008). *EAP Essentials to Principles and Practice.* Lebanon: International Press.

Argent, S., & Alexander, O. (2013). *Access EAP: Frameworks*. Lebanon: International Press.

Biber, D., & Gary, B. (2016). *Grammatical Complexity in Academic English — Linguistic Change in Writing*. Cambridge: Cambridge University Press.

Braine, G. (1995). Writing in the Natural Sciences and Engineering. In D. Belcher, & G. Braine (Eds.), *Academic Writing in a Second Language: Essays on Research and Pedagogy*. New Jersey: Ablex.

Cai, J. G. (2020). *Writing SCI Journal Research Articles for Publication*. Shanghai: Fudan University Press.

Haliday, M.A.K., & Hasan., R. (2001). *Cohesion in English*. Beijing: Foreign Language Teaching and Research Press.

Nesi, H., & Gardner, S. (2012). *Genre Across the Disciplines: Student Writing in Higher Education.* Cambridge: Cambridge University Press.

Swales, J., & Feak, C. (2012). *Academic Writing for Graduate Students: Essential Tasks and Skills* (3rd ed.). Ann Arbor: University of Michigan Press.

Swales, J. M. (1990). *Genre Analysis: English in Academic and Research Settings*. Cambridge: Cambridge University Press.

图书在版编目(CIP)数据

农业学术英语文献阅读教程/姜梅,张吟松主编. —上海: 复旦大学出版社, 2023.9
(复旦卓越规划. 21 世纪大学农业英语系列)
ISBN 978-7-309-16786-3

Ⅰ. ①农… Ⅱ. ①姜… ②张… Ⅲ. ①农业科学-英语-阅读教学-高等学校-教材 Ⅳ. ①S

中国国家版本馆 CIP 数据核字(2023)第 045367 号

农业学术英语文献阅读教程
姜 梅 张吟松 主编
责任编辑/方 君

复旦大学出版社有限公司出版发行
上海市国权路 579 号 邮编: 200433
网址: fupnet@ fudanpress. com http://www. fudanpress. com
门市零售: 86-21-65102580 团体订购: 86-21-65104505
出版部电话: 86-21-65642845
常熟市华顺印刷有限公司

开本 787×1092 1/16 印张 15 字数 352 千
2023 年 9 月第 1 版第 1 次印刷
印数 1—5 100

ISBN 978-7-309-16786-3/S · 18
定价: 48.00 元
